Overleaf: THE MOON MN036 Lunar Farside — I.A.U. Crater No. 308,
diameter 50 miles, 179°E, 5.5°S (Apollo 11) — courtesy of NASA.

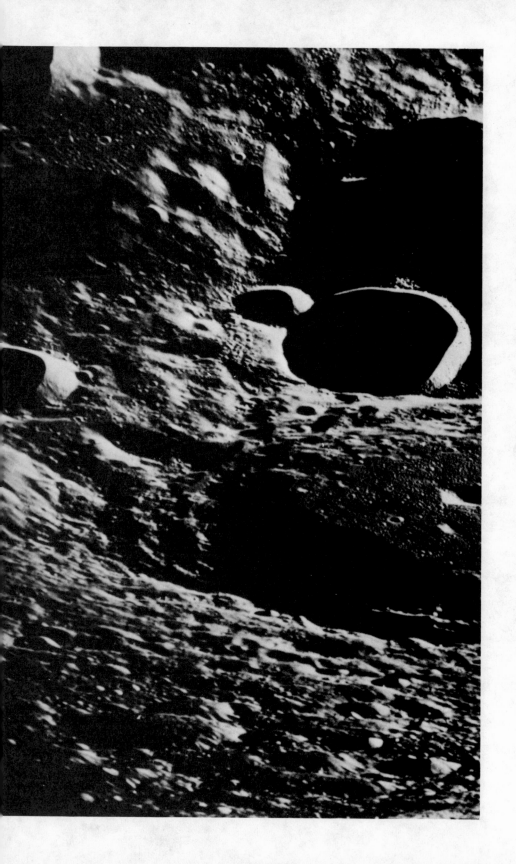

LIFE IN THE UNIVERSE

LIFE IN THE UNIVERSE

Francis Jackson
MB, BS(LOND.), Dip. Bact.(LOND.), FRCPath.

and

Patrick Moore
OBE, DSc(Hon), FRAS

W · W · Norton & Company
New York   London

Copyright © 1987 by Francis Jackson and Patrick Moore
American Edition, 1989.

Printed in the United States of America.

Library of Congress Cataloging-in-Publication Data

Jackson, Francis.
  Life in the universe / Francis Jackson and Patrick Moore — 2nd ed.
  p.   cm.
  Bibliography: p.
  Includes index.
  1. Life on other planets. 2. Life (Biology) I. Moore, Patrick.
II. Title.
QB54.J23 1989
574.999′2—dc19   88–22389

W. W. Norton & Company, Inc., 500 Fifth Avenue,
New York, N. Y. 10110
W. W. Norton & Company Ltd., 37 Great Russell Street,
London WC1B 3NU

CONTENTS

PLATES

FIGURES

PREFACE

Of all the questions facing us, that of life in the Universe is for many one of the most fascinating. How did living things originate? Are we alone in space? – or do other beings exist, possibly far more advanced than ourselves? How many planet-families are there in the Galaxy, and are we likely to find worlds which are similar to the Earth on which we live? Can we ever hope to make contact with extraterrestrial civilizations?

The astronomer is no longer a 'man apart'. His science has become increasingly interwoven with mathematics, physics, chemistry, engineering and – significantly – biology. The search for life in the Universe is now treated seriously as new investigational techniques have been developed. Space medicine has become a study of vital importance, particularly when there is serious talk of establishing permanent colonies beyond the Earth and even of sending astronauts to Mars.

Of the two authors of this book, one (Jackson) is a bacteriologist who has been engaged in research and teaching, and the other (Moore) a lunar and planetary observer. What we have tried to do is to show how the disciplines of biology and astronomy can be usefully combined in a joint approach to the overall question. It has, too, been necessary to make reference to results of recent work in physics (in particular, irreversible thermodynamics), in an attempt to throw new light on the question of life's origin and development.

When the first edition appeared, in 1962, the space age was a mere five years old, and no controlled landings had been made on any other world. Many ideas on planetary conditions held at that time, based largely on observations made from the Earth's surface, have now been proved to be wrong, even with respect to our neighbours Mars and Venus. Conditions on the planets are in many ways quite different from what had been supposed. There have also been significant changes in ideas about conditions on the primitive Earth, including the composition of the atmosphere. Consequently, we have rewritten much of the book and have included a large amount of new information. We hope that you will find the result thought-provoking and enjoyable.

The second author wishes to point out that the biological sections, which make up the major part of the book, have been written entirely by Jackson, so that Moore's rôle has been a comparatively minor one.

F.L.J.
P.M.

LIFE IN THE UNIVERSE

I THE UNIVERSE AROUND US

Life on the Earth can take many forms. Human beings, intellectually the most advanced of the mammals, cannot be regarded as physically hardy but are able to survive under rigorous conditions by virtue of superior intelligence; animals of various kinds exist all over the Earth, from the hot to the cold, wet to dry regions. Some plants and fungi are tolerant of extreme conditions, and micro-organisms are to be found in what at first sight are unlikely habitats, such as hot springs, and even within the surfaces of rocks in the frozen wastes of Antarctica.

Conditions on other worlds close enough to be subjected to detailed examination are different from those on Earth. The Moon, our nearest neighbour in space, is devoid of atmosphere; Venus is intolerably hot, with a crushing atmospheric pressure at its surface, and its clouds contain sulphuric acid; Mars has a very tenuous atmosphere and an extremely cold climate by terrestrial standards, while the remaining planets are even less welcoming. Yet we must remember that our Solar System is but a very small unit of the Universe, and in discussing the problem of life in general, we must take care not to be parochial. First, we must see what the astronomer has to tell us.

Status of the Earth

In ancient times, the Earth was usually considered to be flat, and to lie in the centre of the Universe, with the celestial sphere revolving round it once a day. The earliest astronomers, such as the Chinese and the Egyptians, carried out accurate positional observations of the various objects in the sky but made surprisingly little effort to explain them, and it was only with the advent of the Greek philosophers that astronomy began to take on the characteristics of a

1

true science. The first of the philosophers of the Ionian school was Thales of Miletus, who flourished about 600 BC; the last great astronomer of Classical times, Ptolemy (Claudius Ptolemæus), died about AD 180. During the intervening period, human knowledge increased tremendously but progress was not made at lightning speed. Ptolemy was as remote in time from Thales as we are from the Crusades.

Thales seems to have learned much of his basic astronomy from the Egyptians, and his ideas were naturally primitive by present-day standards. In 544 BC we find Heraclitus of Ephesus maintaining that the diameter of the Sun was no more than about one-third of a metre. The first real step forward was the discovery that the Earth is not flat, but spherical. The great geometer Pythagoras, who was born about 572 BC, was one of the first to realize that the Earth is a globe.

Even in the greatest days of Greece, when Athens was dominant, prejudice was rife. One sufferer from it was Anaxagoras of Clazomenæ, who was born about 500 BC and who (fortunately for himself) enjoyed the friendship and protection of Pericles, the most powerful man in Athens. Anaxagoras maintained that the Sun was a red-hot stone larger than the Peloponnesus, and that the Moon was 'earthy'. For these views he was accused of impiety, and was banished from the city.

The second major step was taken by Aristarchus of Samos (310–230 BC), one of the most brilliant of all the Greek philosophers. By his day, the spherical nature of the Earth was widely accepted but it was also assumed that our world was the centre of the Universe, with the Sun, Moon, planets and stars circling around it. Aristarchus boldly dethroned the Earth from its central position and put forward the *heliocentric* theory, according to which the Sun is in the centre with the Earth moving around it. Unfortunately he could give no conclusive proof, and he found few followers, so that the later Greeks returned to the old *geocentric* (central Earth) picture. Indeed, it was not until the seventeenth century that Aristarchus was finally vindicated.

The scale of the Universe

Though the Greeks (or most of them) believed in the supreme importance of the Earth, they made a start in trying to visualize the size of the Universe. About 270 BC, Aristarchus made the first attempt to measure the distance of the Sun, using a method which was sound enough in theory but inaccurate in practice. He estimated a distance of 4,800,000 kilometres. Eratosthenes of Cyrene (c. 276–196 BC), librarian at Alexandria, estimated the circumference of the Earth with remarkable precision, and the last of the great astronomers of the Greek school, Hipparchus and Ptolemy, left astronomy in a relatively flourishing state: Ptolemy's value for the distance of the Sun was as much as 8,000,000 kilometres. It was known that the Moon and planets were worlds in their own right, and that the stars were much more remote. Plutarch (c. AD 46–120) wrote a curious and apparently incomplete essay entitled *De Facie in Orbe Lunæ*, in which he stated that the Moon is 'cleft with many deep caves and ruptures', and commented that the idea of life there was no more implausible than assuming life to survive in the oceans of the Earth. This was highly significant. To Plutarch, the Moon was an 'earthy' world; he could not know that it lacks atmosphere and water, and there seemed no obvious reason to doubt its habitability. Why should life not appear wherever conditions were suitable for it? The question, still being asked, of whether life will arise wherever conditions are 'suitable' is one to which we shall return later. It is one of those seemingly simple questions which tend to have complex answers.

Ancient astronomy ends abruptly with the death of Ptolemy, whose great book, which has come down to us by way of its Arabic translation (the *Almagest*), provides us with an invaluable summary of the scientific knowledge which had been accumulated in Classical times. There followed a long period of stagnation, but then astronomy began to make further progress – first with the Arabs, who were skilled observers, and then in Europe. In 1546 came the publication of the great book by Copernicus, *De Revolutionibus Orbium Cœlestium*, which cast aside the Ptolemaic theory of the Universe and relocated the Sun to the centre of the Solar System. Later in the sixteenth century, the Danish astronomer Tycho Brahe made precise observations of the positions of the stars and the movements of the planets, thereby enabling his assistant and successor, Johannes Kepler, to establish the heliocentric theory once and for all. In 1608, the telescope was invented, and Galileo and others used it to examine the sky, though the story of what we may term 'modern' astronomy did not begin until the publication of Isaac Newton's immortal *Principia* in 1687.

In 1600, the science of astronomy was still in a more or less mediæval condition, mainly because it was limited to naked-eye studies. The size of the Earth was still a matter for debate (Columbus, on his voyage of discovery, used a value which was less accurate than that of Eratosthenes, which may be why he came home without any clear idea of where he had been), and the official view was still that the Sun moved around the Earth; Galileo was brought to trial and condemned by the Inquisition for daring to teach the Copernican theory. By 1700, all was changed. The

Ptolemaic theory had gone for ever, and the outlook had become essentially modern. Ole Rømer had determined the velocity of light with considerable accuracy, and it was known that the distance of the Sun was at least 138,000,000 kilometres (Cassini, 1672). The Earth had been relegated to planetary status, markedly inferior in size and mass to some other members of the Solar System, such as Jupiter and Saturn.

However, the problem of the distances of the stars proved to be difficult to solve, and defeated even an observer as skilled as William Herschel. Success finally came in 1838, when F. W. Bessel, at Königsberg, used the parallax method to show that a fifth-magnitude star in the constellation of the Swan, 61 Cygni, is roughly 11 light-years away from us (one light-year, the distance light travels in a year, is equal to 9.4607 million million kilometres). Only a few stars are closer than this, and of the stars within a dozen light-years only Alpha Centauri, Sirius, Epsilon Eridani, 61 Cygni, Epsilon Indi, Procyon and Tau Ceti are visible with the naked eye.

As soon as some star-distances became known, the luminosities of the stars concerned could be estimated, and it became clear that the Sun is by no means distinguished. Sirius, which shines as the brightest star in the sky, is 8.6 light-years away and has a luminosity 26 times that of the Sun; Rigel, in Orion, at a distance of approximately 900 light-years, is the equal of about 60,000 Suns, and some stars known to us are more than a million times as luminous as the Sun. On the other hand, we must not feel too humble. Just as the stellar system includes 'searchlights', so it also includes 'glow-worms', and many of the dwarf stars are much less luminous than the Sun.

For our present theme, the most important question is whether other stars may be the centres of planetary systems. The problem will be discussed in detail below, but there are many stars in our galaxy and in other galaxies which are very similar to the Sun, and we have excellent if not conclusive evidence that solar systems are common. Why should the Earth be unique? In a Universe as vast as ours we might reasonably expect to find other not too dissimilar planets, but final proof of their existence will depend on application of sophisticated astronomical techniques, probably in the not too distant future. Yet in our search for extraterrestrial life we must necessarily begin near home, and see what we can learn from the planets of the solar system in which we ourselves live.

The planets and satellites

The Solar System consists of one star – the Sun – the nine planets, and various lesser bodies such as the satellites, comets, asteroids and meteoroids. The Sun itself has a diameter of 1,392,000 kilometres, and a surface temperature of nearly 6000°C; at its centre the temperature must rise to approximately 15 million °C, and it is here that its energy is being produced by nuclear reactions – essentially, the conversion of hydrogen into helium. The Sun, in the generally used system of stellar classification, is a typical Main Sequence star, and is in a stable condition. Not for something of the order of 5000 million years hence will it start to run short of available hydrogen 'fuel'.

The planets are divided into two main groups. The inner group consists of four comparatively small, solid worlds: Mercury, Venus, the Earth and Mars. Beyond Mars we reach a region in which move thousands of worlds which are true midgets, and are known as *asteroids*; even Ceres, the largest of them, is only 1000 kilometres in diameter, and most of the rest are far smaller. Further out we come to the giants: Jupiter, Saturn, Uranus and Neptune, which have gaseous surfaces and are rich in hydrogen. In addition there is Pluto, discovered by Clyde Tombaugh in 1930 as a result of a deliberate search carried out at the Lowell Observatory in Arizona. However, the status of Pluto is very much in doubt. It is very small by planetary standards, and is inferior to our Moon in both size and mass; it also has an exceptionally eccentric path which brings it at times closer-in to the Sun than Neptune. Indeed, between 1979 and 1999 Neptune is unquestionably the outermost known planet.

Several of the planets have families of satellites but most of these are insignificant. Jupiter has four large satellites – Io, Europa, Ganymede and Callisto, collectively called the Galilean satellites, as they were first studied by Galileo. Saturn has one large satellite, Titan, and Neptune also has one, Triton. All the other planetary satellites are below 3000 kilometres in diameter, apart of course from our own Moon, which has 1/81 the mass of the Earth and may lay claim to being a companion planet rather than a mere attendant.

When looking for carbon-based life – the only kind we know with certainty to exist – we must concentrate upon worlds which are neither too hot nor too cold, which have atmospheres containing water vapour and showing evidence of atmospheric modification by biological activity, for example being rich in oxygen, and not in a state of equilibrium. A solid surface with accumulations of liquid water as oceans or lakes would seem to be desirable. Most of the members of our Solar System are unpromising as possible abodes of life. The Moon, Mercury, the asteroids and all the satellites apart from Titan and Triton lack atmospheres of appreciable density, while that of Mars is extremely tenuous, as is that of Pluto, which is also extremely cold. The atmosphere of Venus *is* very dense, but is made up mainly of carbon dioxide, with clouds of sulphuric acid.

The giant planets have no detectable solid surfaces, though probably they do have relatively small silicate cores. In fact, the Earth is the only world in the Solar System well suited to our kind of life. We shall later examine some arguments relating to the possible existence of organisms on other seemingly inhospitable worlds.

Stars and Galaxies

The star-system or Galaxy in which the Sun is situated contains about 100,000 million stars, ranging from the very luminous supergiants through Main Sequence stars down to dim red dwarfs. The Galaxy takes the form of a flattened system with a central bulge, and is about 100,000 light-years in diameter; the greatest thickness is around 20,000 light-years, and the Sun lies near the main plane, approximately 30,000 light-years from the galactic nucleus. We cannot see the centre of the Galaxy, because there is too much 'dust' in the way, but we can detect infra-red radiations from it, and it may well be the site of a very massive black hole. If we could see the Galaxy from 'above' or 'below', it would appear spiral in form, like an immense Catherine-wheel; the sun lies just beyond the edge of one of the spiral arms. The entire Galaxy is rotating. The Sun takes about 225 million years to complete one revolution, a period which is sometimes termed the 'cosmic year'. One cosmic year ago, the most advanced life forms on Earth were reptiles, and the age of the ferocious dinosaurs lay mostly in the future.

The first reasonable picture of the shape of the Galaxy was given by Sir William Herschel, who is usually remembered for his discovery of the planet Uranus in 1781, but whose best work was on stellar astronomy. Herschel also catalogued many of the misty patches which we call nebulæ, and speculated as to whether the so-called 'starry nebulæ' could be independent galaxies in their own right. For many years this suggestion was discounted, but it was proved correct, in our own century, by an indirect method.

Most stars, including the Sun, shine steadily over very long periods. Others show short-term fluctuations in brightness, and are known as *variable stars*. Among these variables are the Cepheids, named after Delta Cephei, the first-discovered member of the class. The Cepheids are absolutely regular in their behaviour, and it was found that the luminosity is linked with the period of variation – the longer the period, or time taken to pass from maximum to maximum, the more luminous the star. It thus became possible to determine the distances of the Cepheids merely by observing their changes in brightness.

In 1923 E. E. Hubble, using the 100-inch Hooker reflector at Mount Wilson in California (then the most powerful telescope in the world) detected Cepheids in several of the 'starry nebulæ', including Messier 31, the Great Spiral in Andromeda, which is dimly visible with the naked eye. Hubble was able to estimate the distances of the Cepheids, and hence of the systems in which they lay. At once the problem was solved: the systems lay at immense distances, and were indeed galaxies. Modern measurements give the distance of Messier 31 as 2,200,000 light-years, and show that it is larger than our Galaxy; even so, it is one of the very closest of the external systems, and is a member of what we call the Local Group.

The Cepheid method can take us out to several tens of millions of lights years but, for more remote systems, even less direct methods have to be used, most of which depend upon spectroscopic evidence. However, we are now confident that galaxies can be detected out to well over 12,000 million light years, so that we are now seeing them as they used to be when the Universe was young – at least in its familiar form. We also know that the entire Universe is expanding, and it is the effect of this expansion on the spectra of radiations reaching us that enables the distances of remote luminous objects to be estimated (see p. 16 Spectroscopy). Our Universe is now generally considered to be finite but unbounded, and the expansion of space itself gives rise to the recession of galaxies (or groups of galaxies) from each other. The more remote an object is from us, the more rapidly it will be found to be receding. This picture – of an expanding Universe which came into existence some 15,000 million years ago in what is colourfully called a 'Big Bang' – is now fairly well established, and we shall return to it later, as it does provide a basis for understanding some important aspects of the problem of the existence of life. In such a Universe, there is no one 'centre', as the Universe will look essentially the same from any viewpoint. We therefore see the old controversies about whether the Sun or the Earth is central in a new perspective. The 'centre' is nowhere and everywhere, which is another way of saying that the question was not entirely appropriate to the type of Universe we now envisage.

We must now turn from the vastness of space and galaxies to consider the very small, and indeed in some respects the inconceivably small, features of the structure of matter, non-living and living.

Materials of the Universe

As is well known, the materials we find around us in the world have been shown to consist of myriads of exceedingly minute *atoms*. Substances which consist entirely of one kind of atom are called *elements*, and although atoms of a particular element cannot be changed into those of another by chemical reactions, transformations can be achieved by physical methods. Elements termed *radioactive* consist of unstable atoms and, as will be described later, these change spontaneously. Atoms of the same or different elements frequently unite to form *molecules*, and substances whose molecules contain atoms of two or more different elements are called *compounds*. Ninety-two naturally-occurring elements are known, and others have been produced artificially.

Following the pioneer work of J. J. Thomson, Max Planck, Ernest Rutherford and Niels Bohr, it was found useful to picture an atom as consisting of a central *nucleus* bearing a positive electric charge, with negatively charged electrons revolving around it in circular orbits, rather like a miniature solar system. This picture was not entirely satisfactory but it made possible a better understanding of the nature of simple atoms. The work of Heisenberg, leading to the development of quantum mechanics, Louis de Broglie's recognition that electrons and other elementary particles should exhibit wave-like characteristics, and Schrödinger's theoretical work on wave mechanics, made possible the formulation of a new mathematical expression of atomic structure. Electrons can be considered as vibrating electric charge, and this sets certain constraints on the energy states in which atoms can exist. Electrons around atomic nuclei produce *standing waves* which define the space within which the electron as a particle can be considered to be in rapid motion. The notion that matter has wave-like as well as particle-like features is expressed in the concept of the *wave-particle duality*, recognizing that, depending on the particular experimental situation, electrons and other constituents of matter may give rise to phenomena interpretable as due to wave-like properties or to particles. This does not mean that an electron is 'really' a 'particle' and a 'wave' but rather that the phenomena can be interpreted by the mathematical methods appropriate to such concepts. To use the term introduced by Niels Bohr, the concepts of wave and particle are *complementary*, and will manifest under appropriate conditions.

Although the notion of electrons in atoms as minute hard particles in circular orbits has been abandoned, it is convenient and sufficient for many purposes to consider atoms as consisting of a nucleus surrounded by orbiting electrons moving at high speed. The electrons are arranged in *orbitals*, an orbital being visualized as a *charge cloud* which is not uniform in density. The greater the *electron density* of the 'cloud', the greater the *probability* of the electron being found in that region, and the probabilities can be calculated with precision (using quantum mechanics). Atomic structure follows certain rules, as only some arrangements of the constituents of atoms are possible or stable. The Pauli principle states that any one orbital cannot contain more than two electrons, and orbitals are arranged in sets according to their *principal quantum number*, *n*, a set of orbitals having the same principal quantum number *n* constituting a *shell* or *quantum level* in the atom. In addition there is an *auxiliary* quantum number, *1*, relating to the geometrical form of orbitals, a *magnetic quantum number*, *m*, and finally a quantum number, *s*, indicating a property known as the *spin* of individual electrons. The chemical properties of an atom are strongly determined by the arrangement of its outermost electrons, the level at which interaction with other atoms takes place, and the number of electrons is itself a reflection of the number of protons in the nucleus.

For our present purposes, it will be sufficient to consider atomic nuclei as composed of *protons* and *neutrons*, (collectively called *nucleons*) and we shall not discuss current views on the constitution of these as aggregations of more fundamental entities called *quarks*. A proton bears a positive electric charge, and a neutron is electrically neutral. Each of these has a mass approximately 1836 times the mass of an electron. The mass of a proton is the fundamental unit of atomic mass.

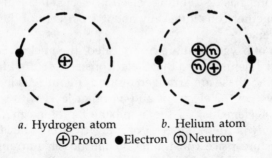

a. Hydrogen atom   b. Helium atom
⊕ Proton ● Electron ⓝ Neutron

Fig. 1 Hydrogen and helium atoms

The simplest atom is that of hydrogen. An atom of hydrogen consists of a single proton and one electron (Fig. 1a) and like all complete atoms is electrically neutral, because the positive charge of the proton is balanced by the negative charge of the electron. The atom of common hydrogen contains no neutrons, but the nuclei of all other atoms contain a number of neutrons, which affect the atomic mass (often inappropriately called the atomic *weight*) but not the nuclear charge. In all atoms other than that of hydrogen, there are at least as many neutrons as protons, and usually more, and they are necessary for the stability of the larger nuclei. If an atom loses one or more of its surrounding electrons, it becomes positively charged, because there will then be an excess of protons; conversely, an atom which gains one or more electrons will be negatively charged. Charged atoms are called *ions*.

Charges of like sign repel each other and the nuclei of atoms, with the exception of hydrogen, contain a number of positively charged protons (together with the neutrons). The repulsive effect of these electric charges is counteracted by a strong nuclear force, the *strong interaction*, which has a very short range. Neutrons are not subject to the electrical repulsion effects of the positive charges but are subject to the strong interaction. Because of its short range, this force affects only particles that are very close to each other, and this has implications for the structure of nuclei. Moreover, protons and neutrons are subject to the Pauli principle, so that the nucleus itself has a 'shell' structure. These points are mentioned to emphasize that there are important constraints on the forms that atoms can take, and that atomic structure is confined to certain patterns.

Hydrogen is the most abundant element, accounting for about 93 per cent of the total number of atoms in the Universe and 76 per cent of the total mass. The next most common element is helium, about 7 per cent by number of atoms and 23 per cent by mass. The atom of helium contains 2 protons and 2 neutrons in its nucleus, and therefore has 2 orbiting electrons (Fig. 1b). All the other elements together constitute only about 1 per cent of the total mass of the Universe.

The atom of oxygen is more complicated, the nucleus containing 8 protons and 8 neutrons, so that 8 electrons are required for electrical neutrality, and these are arranged around the nucleus in two shells, an inner level with 2 electrons and an outer level with 6 (Fig. 2). In the series of elements, as the number of protons in the nucleus increases, so will the number of orbiting electrons, but the rules of atomic structure place constraints on the possible arrangements of electrons in successive shells. The work of Max Planck, Einstein and others showed that energy exchanges between radiation and atoms, or *vice versa*, do not occur in a quantitatively continuous manner but in 'packets' whose size is a multiple of a unit of energy. Moreover, for each frequency of radiation, the unit or *quantum* has a specific value, given by the expression $E = h\nu$, in which E is the energy, ν the frequency and h is Planck's constant (6.6×10^{-34} Joules sec.), and these quanta of energy can exhibit particle-like qualities. Quanta of radiation are called *photons*. When a photon of appropriate energy is absorbed by an atom, an electron may be shifted to a higher energy level. Subsequently, when the electron shifts back to the original level, a quantum of energy is emitted.

The shells or quantum levels surrounding the nucleus of the atom are numbered 1, 2, 3, 4, . . . etc., from within outwards, the number assigned being the *principal quantum number*. The capacity of the

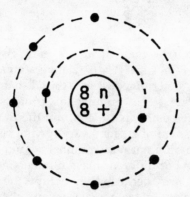

Fig. 2 The oxygen atom

different shells for electrons is not the same, and the greater the radius of the shell, the more electrons it can contain. Thus, the first shell can accommodate at most 2 electrons, the second 8, the third 18 and the fourth 32. The restriction on the numbers of electrons per shell prevents the atom collapsing until all of its electrons are close to the nucleus.

The number of protons in the nucleus is the *atomic number* and indicates the nuclear charge, and in electrically neutral atoms the number of orbiting electrons is the same as the atomic number. The number of neutrons in the nucleus is the *neutron number* and the sum of the proton number and neutron number gives the *mass number*. For the expression of relative atomic masses, the unit chosen is one-twelfth of the mass of the atom of the commonest *isotope* (see below) of carbon.

We can imagine the series of increasingly complex atoms being built up in an orderly fashion from the necessary components. Hydrogen has 1 electron in the first shell and helium 2, and no more than 2 electrons can be present in this shell. The next element, lithium, has 2 electrons in the first shell and 1 in the third. With two exceptions, the outermost shell of the atom of an element does not contain more than 8 electrons. When a shell is occupied by 8 electrons, no more can be added until the next shell out has been occupied. An atom with a complete *octet* (= 8) of electrons in its outermost shell is usually chemically unreactive, because of the inability of the stable octet of electrons to interact with the outermost electron shells of other atoms. The noble gases, neon, argon, krypton and xenon have stable octets in shells 2, 3, 4 and 5 respectively. Helium is also chemically inert, because the first shell, containing 2 electrons, is already full.

Isotopes

It was at first thought that all atoms of a given element were identical, but this is not so. When relative atomic masses of many elements had been determined, it was found that they were usually not whole numbers. The number of protons in the nucleus determines which element an atom is, but the number of neutrons, and hence the mass number, may vary, giving rise to atoms of the same element with different mass numbers. Atoms having the same number of protons but different mass numbers are termed *isotopes*, (Greek for 'same place') because they occupy the same place in the *periodic table* (see below) of elements. They are chemically similar but physically different. Relative atomic masses which are not whole numbers reflect the mixture of different isotopes occurring under natural conditions, and are mean values.

The common hydrogen atom consists of 1 proton and 1 electron, but there are atoms with 1 proton, 1 neutron and 1 electron (deuterium) and 1 proton, 2 neutrons and 1 electron (tritium). Deuterium and tritium are therefore isotopes of hydrogen. The deuterium atom is stable, but the tritium nucleus 'decays' spontaneously by emitting an electron, so converting one of the neutrons to a proton. The resulting nucleus contains 2 protons and 1 neutron, and is an isotope of helium (number of protons is 2). Hydrogen, deuterium and tritium have similar chemical properties but the atom of common hydrogen, because of its small dimensions, is in some ways a special case. Many other elements have isotopes, radioactive and non-radioactive, and some isotopes have important applications in industry, medicine and biological research.

As long ago as 1869, the Russian chemist Mendeleev, elaborating on earlier suggestions by Newlands and Döbereiner, proposed a *periodic table* of elements. He noticed that if a list were made of elements in order of atomic masses, there was a periodic change of properties of the elements. Moreover, it was apparent that the table which could be constructed at that time was incomplete, and Mendeleev was able to forecast, with remarkable accuracy, the main properties of some elements which had not then been discovered. Later work extended and modified the periodic table and it is now clear that the regularities noticed by Mendeleev have a sound basis in atomic structure. Recognition that chemical properties are more closely related to atomic number than to atomic mass has thrown light on some apparent anomalies of earlier versions of the table.

Uncertainty in the world of physics

When physicists are dealing with large objects of the type with which we are familiar in everyday life, it is possible to measure simultaneously the location and momentum (mass × velocity) of an object with high precision. With entities of the nature of isolated atoms or subatomic particles, this is not so. As Heisenberg argued, in order to define these features radiation of suitable wavelength must be used, and for very small entities, the wavelength of the radiation necessary to make the measurement must be very short, if the determination of position is to be precise. The energy of the photons constituting this 'probe' will disrupt the motion of the particle and render precise determination of the momentum impossible. Using photons of lower energy, and hence longer wavelength, will reduce the disruption of motion but lead to loss of information on position. Heisenberg's principle of uncertainty states that the product of the uncertainties of the measurements of position and momentum can never be less than Planck's constant (h). Precise prediction of the future state of any group of particles is therefore in principle impossible, although the error will be negligible with large objects. In the subatomic world, indeterminacy is a major factor, and mathematical treatment of events has to be statistical, as in quantum mechanics. At the macroscopic level, for everyday purposes, determinism rules. Strictly speaking, all measurements lack exact precision, and the act of probing the Universe for information itself changes the state of the Universe. Heisenberg postulated that what we call the 'path' of an object comes into existence only when it is 'observed'. Wheeler (1982) has described the Universe as a 'self-excited circuit', very small at the time of the Big Bang, growing in size with expansion and giving rise to life, observers and observing equipment which itself, through elementary quantum processes terminating on it, plays a rôle in giving 'reality' to events.

In classical physics, the observer was visualized as separate from the observed, but in quantum physics the observer is inevitably promoted to a rôle as 'participator'. In Wheeler's words, 'in the real world of quantum physics, *no elementary phenomenon is a phenomenon until it is a recorded phenomenon*'. When a photon is intercepted so that the photon acts at a particular location, as by producing change in a photographic film or the retina of an eye, the destination of the photon, and so its 'path', has been fixed. Before this event, involving a change of state of the system, the photon was not like a particle travelling through space on a precise path, but was potentially localizable by interception by an 'observer'. In Wheeler's view, consciousness has nothing to do with the quantum process, the

13

important thing being the registering of an event by an irreversible act of amplification. He leaves open the possibility that this could be the first step in the translation of measurement into meaning, by entry of the record into the consciousness of a person, animal or even a computer, but he regards this as separate from quantum phenomena.

Eugene Wigner (1982), in a discussion of the limitations of the validity of present-day physics, emphasizes that the basic principles of quantum mechanics are fundamentally different from those of earlier theories, and that quantum mechanics represents a radical departure from classical concepts. Wigner has stated that present-day physics and chemistry may lose validity with respect to the more developed animals with minds and consciousness and that quantum mechanics cannot describe the impressions of a 'truly living person'. It is also possible, he argues, that quantum mechanics does not give an adequate description of measuring apparatus, but the possibility that some attributes of living organisms may be outside present-day physics does not rule out the future development of physics to include them. Electricity and magnetism were outside Newton's physics, and only recently have subatomic events been incorporated into the body of physical science. (It should be noted that there is much disagreement on this and related matters. For a realistic and common-sense interpretation of quantum theory, see *Quantum Theory and the Schism in Physics*, Karl Popper, 1982. This contains his paper 'Quantum Mechanics without the Observer'.)

The indelible registration of events, as described by Wheeler, is regarded as giving 'tangible reality' to the Universe, and in addition it would seem that consciousness and intelligence give meaning and value. Organisms are part of the Universe, and it seems to be in principle impossible to separate the observer from what is observed. Viewed in this way, the rôle of life in the Universe appears to have greater significance than was suggested by the older nineteenth- and early twentieth-century materialistic and behaviouristic outlooks. One might argue that the significance of life is awe-inspiring, as it is playing a creative role, and the possibility that organisms might be present on only a few planets in the entire Universe, or even only on one, would not alter this evaluation. Life as we know it might prove to be a transition stage to something more enduring and profound, and perhaps consciousness and intelligence could be generated in non-carbon based systems, even in complex computers that appear to lie in our not too distant future. Already, carbon-based life has produced the silicon chip, a form of organized matter that could almost certainly never have sprung directly and unaided from a primordial mix of simple chemicals. There may well be aspects of the Universe yet to be discovered of which we are as ignorant, and perhaps at present as incapable, of understanding, as the average ant presumably is of mastering the theory of general relativity.

Considerations of this kind are having important effects on the development of ideas in theoretical physics, information theory and related subjects. It has become apparent that our modes of thinking about the Universe at the scale we perceive it with our unaided senses are inappropriate for application at the atomic and subatomic levels. Recently, experimental evidence has been obtained that the behaviour of matter in the Universe may also be influenced by some coordinating process that lies outside our ordinary experience. As Hubert Reeves (1985) has expressed it,

> there might be two levels of contact between things. First would be the level of traditional causality. Second would be a level that involved neither the force of one body on another, nor any exchange of energy. Rather, it would seem to involve an immanent and omnipresent influence that is difficult to describe with precision. I would very much like to know what connections might exist between this hypothetical influence and cosmic evolution.

Formation of elements

The production of the nuclei of atoms is believed to have begun during the first few seconds of the explosion, the 'Big Bang', initiating the formation of our Universe. At this time, quarks are presumed to have associated to give rise to protons and neutrons, so that hydrogen nuclei (protons) would have been present. Within a few seconds, helium nuclei were formed, and some lithium-7 nuclei. The heavier nuclei are synthesized inside stars, 'cooked' in the stellar interiors, and are released by supernovæ explosions. The formation of helium from hydrogen is also a process of fundamental importance inside stars, as a source of energy which is released as gamma rays when nucleons aggregate to form nuclei. Because of the equivalence of energy and mass ($E=mc^2$), the energy given off during fusion carries away some mass, and a nucleus consisting of protons and neutrons has a mass somewhat smaller than the sum of the individual masses of the same number of uncombined nucleons. In the cores of red giants, helium is transformed into carbon and oxygen, and in suitable stellar environments, elements including neon, sodium, magnesium, aluminium, silicon and phosphorus are produced, and other elements are generated before the supernova explosion. The appropriate temperatures for promotion of nuclear syntheses are provided at various levels in stars of different ages. In

15

a sense, then, we are children of the stars: most elements, including those of biological significance, were born in stellar processes.

Radioactivity

The radiations emitted by radioactive atoms arise from the nuclei. Radium was one of the first radioactive elements to be investigated, and three different kinds of 'rays', termed by Rutherford alpha-, beta- and gamma-rays were distinguished. The alpha- and beta-rays were deflected by magnetic fields, and from the direction of the deflection it was concluded that alpha-rays were positively charged and beta-rays negatively charged. Later, it was shown that alpha-rays are streams of particles indistinguishable from the nuclei of helium atoms, while beta-rays are streams of electrons. Gamma-rays are short-wave electromagnetic radiations of the same kind as X-rays and light.

The emission of these 'rays' is associated with changes in atomic nuclear constitution. The expulsion of an alpha-particle, which in atomic terms has a mass of 4 and a proton number of 2, lowers the atomic mass by 4 and the proton number by 2. The expulsion of a beta-particle, an electron arising from the transformation of a neutron into a proton, an electron and a third particle, an *antineutrino*, leaves the atom with one more proton than it had before, so raising the proton number by 1 without the atomic mass being appreciably affected, because the mass of the expelled electron is so minute.

In this brief account only a few aspects of atomic structure have been dealt with, to provide a basis for the understanding of later chapters. Many other particles besides those mentioned are known, and some play a rôle in holding the nuclear components together. There are also 'antiparticles', which correspond to most known types, with opposite characteristics to those commonly encountered. Thus the 'antiproton' is negatively charged but has the same mass as the proton; the *positron*, the antiparticle of the electron, is similar to an electron but is positively charged. This raises the possibility that 'antimatter', constructed from antiparticles, might exist in some regions of the Universe. Our kind of matter and antimatter if brought into contact would mutually annihilate in a burst of radiation.

Spectroscopy

White light is a mixture of electromagnetic radiations of different wavelengths, and a beam of white light can be split into its coloured components by means of a spectroscope. The *spectrum* of visible light extends from violet radiations, with wavelengths of about 4×10^{-7} m through blue, green, yellow, orange, and finally red with wavelengths around 7.5×10^{-7} m. Radiations in the ultra-violet, X-ray and gamma-ray regions of the electromagnetic spectrum have wavelengths shorter than those of visible light, while the wavelengths of the infra-red (heat) and radio regions are longer. By means of suitable methods, the emission and absorption of radiations, the various zones of the electromagnetic spectrum can be studied, and may give important information about the composition of galactic, stellar, planetary, cometary and interstellar matter.

An incandescent solid gives out a continuous spectrum of radiation in which the colours revealed by an optical spectroscope form a rainbow-like band, but if a glowing gas or vapour is examined spectroscopically, a discontinuous spectrum consisting of a number of bright lines of light of certain wavelengths is produced. These lines are emitted as a result of excitation of the atoms of the substance and their position in the spectrum, which can be measured with great precision, is characteristic for each particular element. The radiations are emitted when electrons in excited atoms shift back to a lower energy level. A spectrum of this type is called an *emission spectrum*.

If light from some source giving a continuous spectrum is passed through a gas or vapour, the atoms absorb radiation of the same wavelength they would emit under the conditions described above, so that examination of the radiations that have passed through the gas will show deficiencies, appearing as dark lines, in the continuous spectrum. The positions of these lines is characteristic of the elements which have absorbed the radiation, and form what is known as an *absorption spectrum*.

Spectroscopy can also be applied to the study and detection of molecules of compounds, and may give valuable information. Extension of the range to include ultra-violet and infra-red radiation will give more information than examination of the visible range alone. In recent years, radioastronomy, working at longer wavelengths, has given important new information on molecules present in space. Indeed, we can see that the radiations reaching us from extraterrestrial sources contain a great deal of information and they have revealed much about the composition of stellar atmospheres, temperatures and stellar and galactic motions, the last because relative motion of the source of radiations and the receiver will affect the position of lines in the spectrum. Recession of source and receiver causes increase of wavelength (Doppler effect), a *red shift* of the spectral lines characteristic of elements, and the degree of shift is related to the velocity of recession. It is because data derived from the study of galaxies is consistent with predictions of the theory of an expanding universe that that theory is currently widely accepted as probably essentially correct.

The Sun is the only star close enough to be examined in great detail, but solar spectroscopy has been invaluable in pointing to conditions of other stars. The situation is less straightforward with the Moon and planets, which shine by reflection of sunlight, but if the target planet has an atmosphere signs of it should be discernible in the spectrum. Before the age of space research, virtually all our knowledge of the make-up of planetary atmospheres was derived from spectroscopic analysis carried out from the surface of the Earth.

Spectroscopy in its various forms has been one of the most valuable techniques available to astronomers, physicists, chemists and biologists.

II THE NATURE AND ORIGIN OF LIVING ORGANISMS

Historical introduction

The problems of the nature and origin of living organisms on the Earth have stimulated speculation and argument for thousands of years, and we can hardly doubt that the striking differences between living and dead animals and human beings profoundly impressed prehistoric peoples. First-hand experience of human existence must have been responsible for the emergence of many early ideas on the nature of living things. It was necessary to account not only for the behaviour of active humans and animals pursuing their everyday lives but also for the seemingly mysterious phenomena of sleep, dreams, disease and death.

It is possible that dreams about dead persons played an important part in the development of the notion of immortality and of the related belief that the living body is inhabited and rendered alive by a shade, ghost, anima or soul which can, under certain circumstances, lead a separate existence, leaving the body temporarily or permanently. This belief has persisted into modern times, and phenomena considered to be examples of 'out-of-body experiences' are being reported and studied up to the present day. It has frequently been suggested that the non-corporeal component of an organism might successively inhabit different bodies, animal, human and even plants, as in the concept of 'metempsychosis', a feature of the Pythagorean School of Greek philosophers, who probably derived it from the Orphic religion. Belief in transmigration of souls was important in ancient Egypt, and is a feature of Buddhism and of the religions of some existing primitive peoples.

To many ancient observers, it seemed self-evident that even complicated creatures, such as worms and birds, could arise from

19

the simpler earth and dirt. Mud might generate mice, and maggots emerge from stale meat. We can distinguish here two types of substance thought to be capable of giving rise to living organisms: first, inorganic materials and secondly materials which, although dead, had recently been living.

The development of organized agriculture and irrigation techniques probably influenced thought on the nature of life by broadening human direct and detailed experience of living things. Crowther (1955) has pointed out that the Ionian Greek philosophers, in Asia Minor, may have derived a fairly extensive and objective view of living things from the agricultural Sumerians of Mesopotamia, an outlook less exclusively associated with men and animals than were the earlier ideas of more primitive peoples.

Greek thought

There was much Greek speculation on the nature and origin of living things, but concerning origins the later Greek philosophers added little of importance to the ideas of the Ionians. Anaximander, a Greek Ionian philosopher who lived in the sixth century BC, came to the conclusion that living organisms were subject to evolutionary change. He suggested that the motions of a fundamental substance had produced the Sun, planets and stars, and that this same substance had given rise to living things which survived by adaptation to their environment, and that man had evolved from a fish-like ancestor. Anaximander believed that the action of the Sun's rays on moist materials had promoted the origin of organisms, and we shall see later that a similar view plays a part in modern theories. These opinions of Anaximander were, when propounded, far in advance of the science of the time, and so could find no substantiation in fact.

Not all philosophers were as far-seeing, or as fortunate in their guesses, as Anaximander. Oparin (1957) quotes the suggestion, attributed to Empedocles (fifth century BC) that from the Earth 'many foreheads without necks sprang forth, and arms wandered unattached, bereft of shoulders, and eyes strayed about alone needing brows'. Complete bodies, animal and human, he supposed to have been formed by the later union of individually produced members. In this somewhat bizarre form, the ideas of Empedocles seem improbable in the extreme, but in modern times there has been recognition of the probable evolutionary importance of the union of genetic material derived from different organisms, and even of the production of new types of organism by union and symbiosis of cells of different species (see for example Margulis, 1981).

Some early thinkers reached the conclusion that there must be a limit to the divisibility of matter. The development of this theory is particularly associated with Democritus (born about 470 BC), who derived the notion from Leucippus (early fifth century BC). Living things were supposed to contain a fluid matter of a special kind, made of finer and more mobile ultimate particles, or 'atoms', than those of ordinary solid objects. Much early Greek thought was essentially, if subtly, materialistic, and this is typified by the view that the animating principle was a fine, mobile fluid. The idea of an immaterial spirit or soul was a later development, for the shades envisaged by more primitive peoples had material, if sometimes magical, attributes, and could be appeased by offerings of food and material objects.

Plato (427–347 BC) and Aristotle (385–322 BC) ascribed the various functions of living bodies to the 'psyche', and distinguished vegetable, animal and intellectual functions. Plato believed that the human psyche was largely independent of the body and capable of a separate existence. Aristotle seems to have regarded the psyche and body ('soma') as inseparable, although he suggested that reasoning might be a manifestation of universal Reason acting through the individual psyche. The works of Plato and Aristotle profoundly influenced the development of Christianity during the next 1500 years, and at the same time the notion of wholly non-material souls grew. It is interesting to note, however, that even some of the Christian fathers did not think of the soul as completely without material qualities, for it was suggested that a non-material soul would not have been able to suffer the torments of Hell, which would have made Hell rather pointless.

We have mentioned that the belief in spontaneous generation of living things from non-living materials was widespread in the ancient world. Aristotle accepted as fact the spontaneous generation of animals and plants from earth, slime and manure, and his opinions carried great weight for many hundreds of years. It was not until the seventeenth century that careful testing of these beliefs began.

Experimental approach

The first experiments which gave clear results leading to a serious questioning of the belief in spontaneous generation were conducted by an Italian, Francesco Redi. In 1668, Redi showed that maggots did not appear on stale meat if the meat was adequately protected from flies; but he continued to believe in some other supposed examples of spontaneous generation. A contemporary of Redi was the Dutchman, Anthony van Leeuwenhoek, the first true microbiologist. The methods and findings of this remarkable man have been brilliantly described by Dobell, in his book *Anthony van Leeuwenhoek and his Little Animals*. A master of the art of lens making, he discovered with his simple microscopes the world of micro-organisms, and reported his findings to the Royal Society, London, in a series of letters. Leeuwenhoek's work led to the recognition of microbes in putrefying materials, and arguments raged about the origin of these tiny creatures. Did they come from the air, or were they generated spontaneously in gravies, soups or other fluids in which they were found?

In the mid-eighteenth century, Spallanzani, another Italian investigator, showed that no microbes appeared in broths subjected to prolonged heating in containers which were quickly sealed before cooling. These findings were challenged by Needham, in England, on the grounds that heating might destroy a 'vegetative force' which resided in the materials used in the experiments. Spallanzani planned and executed a brilliant series of experiments to prove that this was not so. He was able to show, too, that some of the minute creatures that grew and multiplied in soups and broths were able to carry on their activities after removal of air by means of a vacuum pump. 'How wonderful this is,' he wrote, 'for we have always believed there is no living being that can live without the advantages air offers it.'

The nineteenth century

Spallanzani died in 1799, at the beginning of a century of rapid advance in biological knowledge. In 1837, Cagniard de la Tour, a French investigator, found that the fermentation of beer resulted from the activities of tiny organisms. Microscopical examination of fermenting fluids showed that the yeasts present were growing and multiplying by budding. Schwann, in Germany, demonstrated that putrefaction of meat required the presence of microscopic creatures, and that if these were rigidly excluded meat would remain fresh for long periods of time.

The theory that complicated microbes could be generated in soups and broths in a few hours or days received another setback from the work of the great French scientist, Louis Pasteur. He repeated, with various technical refinements, some of Spallanzani's experiments and demonstrated that flasks of boiled broths remained sterile unless contaminated by dust from the air. Germs were not developing spontaneously from broths but appeared only if unfiltered air could enter the flasks. There was some criticism of Pasteur's results by other investigators who claimed that heating did not always prevent the appearance of micro-organisms in soups and infusions. Pasteur used boiled yeast broths, but Pouchet and his colleagues tried infusions of hay and found that organisms appeared sometimes even after heating. The reason for this was discovered later by the English scientist John Tyndall, who showed that hay contains microscopic germs in the form of *spores* which are able to withstand long periods of boiling in water and require more drastic treatment to kill them.

The work of Spallanzani, Pasteur and Tyndall disposed of the naïve theory of spontaneous generation of complex microbes but in no way ruled out the possibility that in the course of hundreds or thousands of millions of years living organisms developed from relatively simple chemical precursors by a long process of chemical evolution. Tyndall was, in fact, an outstanding champion of this view.

Later developments

During the later part of the nineteenth century and the early twentieth century, discussion of the origin of life led to considerable clarification of the problem. As a result of the studies and writings of Charles Darwin, T. H. Huxley, Tyndall, Schäfer and others, a position was reached which has much in common with present-day views.

Tyndall (1876), in his essay on 'Vitality' (1865), summarized the state of knowledge at that time. He recognized that the energy for living was derived ultimately from the Sun, and that plants were essential for trapping solar energy and played an intermediate role in the energy chain from Sun to animals. He emphasized that 'the matter of the animal body is that of inorganic nature. There is no substance in the animal tissues which is not primarily derived from the rocks, the water and the air.' He proceeded to argue that since every portion of an animal body may be reduced to inorganic matter, 'a perfect reversal of this process of reduction would carry us from the inorganic to the organic, and such a reversal is at least conceivable'. Tyndall concluded that it was the special arrangement of elements in living bodies which led to the phenomena of life, and turning to the problem of life's origin he proclaimed, 'In an amorphous drop of water lie latent all the marvels of crystalline force; and who will set limits to the possible play of molecules in a cooling planet?'

It is apparent that this view differs from those earlier theories that postulated a special material principle, of a subtle and tenuous kind, as an essential constituent of living things. The elements of living bodies are not different from those of inorganic nature but have a special arrangement. How this special arrangement might have come about is a question that has exercised the minds of many eminent scientists and still awaits a definitive answer. In his address to the British Association in Belfast (1874) Tyndall, with true nineteenth-century optimism, said,

> the vision of the mind authoritatively supplements the vision of the eye. By a necessity engendered and justified by science I cross the boundary of the experimental evidence and discern in matter which we, in our ignorance of its latent powers, and not-withstanding our professed reverence for its Creator, have previously covered with opprobrium, the promise and potency of all terrestrial life.

Charles Darwin, who was concerned more particularly with the later evolution of living things, gave some thought to possible processes of life's origin. In a remarkable paragraph, he wrote,

> It is often said that all the conditions for the first production of a living organism are present, which could ever have been present. But if (and Oh! what a big if!) we could conceive in some warm little pond, with all sorts of ammonia and phosphoric salts, light, heat, electricity, etc., present, that a protein compound was chemically formed ready to undergo still more complex changes, at the present day such matter would be instantly devoured or absorbed, which would not have been the case before living creatures were formed.

There are indications that Darwin did not consider the subject one for serious scientific consideration. In a letter to Sir J. Hooker he wrote, 'It is mere rubbish thinking at present of the origin of life; one might as well think of the origin of matter.' Bernal (1951) remarked that we are now almost in a position to take Darwin at his word, for the mode of origin of matter in the forms in which we find it on this Earth is at last becoming clearer. Bernal warned that we should not accept 'wild hypotheses of the origin of life or of matter – we should attempt almost from the outset to produce careful and logical sequences in which we can hope to demonstrate that certain stages must have preceded certain others, and from these partial sequences gradually build up one coherent story'. During the past three decades, several workers have tried to follow this advice, and a number of 'partial sequences', with some supporting experimental evidence, have been proposed.

We have seen earlier how Tyndall sought to clarify the problem of the nature and origin of living things. At about the same time, Thomas Henry Huxley published (1868) his discourse *On the Physical Basis of Life*. Tyndall, he pointed out that plants can build themselves up from simple materials, whereas animals require more complicated foods. He argued that, in spite of differences in detail, there is a fundamental similarity between living things, 'a single physical basis of life underlying all the diversities of vital existence'. He stressed that

> the existence of the matter of life depends on the pre-existence of certain compounds; namely, carbonic acid, water and certain nitrogenous bodies. . . . They are as necessary to the protoplasm of the plant as the protoplasm of the plant is to that of the animal. Carbon, hydrogen, oxygen and nitrogen are all lifeless bodies. Of these, carbon and oxygen unite, in certain proportions and under certain conditions, to give rise to carbonic acid; hydrogen and oxygen produce water; nitrogen and other elements give rise to nitrogenous salts. These new compounds, like the elementary bodies of which they are composed, are lifeless. But when they are brought together, under certain conditions, they give rise to the still more complex body, protoplasm, and this protoplasm exhibits the phenomena of life.

Further, he said (our italics), '*I see no break in this series of steps in molecular complication, and I am unable to understand why the language which is applicable to any one term of the series may not be used for any of the others.*'

In the nineteenth century there was still a widespread belief that 'Life' was an entity able to use 'matter', but capable of independent existence. Anyone who questioned this view was exposed to the risk of heavy criticism from religious authorities. Huxley and Tyndall stoutly defended themselves when challenged. At least they could no longer be burned at the stake, as was Giordano Bruno, for expressing unorthodox views, or summoned before priests and subjected to threats and pressures, as was Galileo. Huxley quoted with approval some words of the philosopher David Hume: 'If we take in our hand any volume of divinity, or school metaphysics, for instance, let us ask, Does it contain any abstract reasoning concerning quantity or number? No. Does it contain any experimental reasoning concerning matter of fact and existence? No. Commit it then to the flames; for it can contain nothing but sophistry and illusion.'

Recent attitudes to the origin of living organisms have been influenced by a number of different developments in scientific thought. The rise of molecular biology has led to a more profound understanding of the nature and functioning of organisms, and the mechanisms by which vital processes are controlled. Information theory and molecular genetics have contributed to ideas on how matter might come to be organized in stable complex patterns. Factors contributing to the emergence of what we regard as Order from a seemingly more chaotic state have received increasing attention (see Crutchfield *et al.*, 1986). Astronomical advances have clarified ideas on some aspects of the possibility of finding life on other planets in our Solar System, for the most part negatively, but at the same time have demonstrated that possible precursor molecules for forming the building blocks of living organisms may be widespread in the Universe. The widespread optimism about the chances of finding life on other planets has been tempered by experience, and some leading authorities have expressed the view that the chances of the emergence of life seem so small, in the absence of some cosmic organizing process as yet not understood, that such a directing influence must exist, or ours may well be the only inhabited planet in the Galaxy or even in the Universe. Moreover, developments during this century in fundamental physics have led to a realization that matter itself is difficult to characterize, and that our ignorance of the ultimate nature of the Universe is profound.

The relevance of quantum theory to the understanding of living organisms is uncertain, but it seems likely that it will be of great importance. Bernal (1951) wrote: 'The key to the understanding of the chemical evolution of life lies in the junction between observational biochemistry on the one hand and quantum theory on the other', suggesting that the finding of the key might be a formidable task. Perhaps it is not unreasonable to suggest that recent developments in the understanding of the constitution of matter might in the end make it easier than it was in Tyndall's day to discern in matter 'the promise and potency of all terrestrial life'. Some aspects of life, and in particular the emergence of consciousness, still appear to pose extraordinarily difficult fundamental questions. Are conscious beings indeed the Universe's own sensors and analysers for viewing or enjoying itself? On a more mundane scale, can we regard the entire biosphere of a planet, or indeed the *whole planet*, as in effect a single, self-regulating organism, as has been suggested in the 'Gaia' hypothesis (see Lovelock, 1979, 1986)? These are fascinating thoughts and we may reasonably suspect that answers will ultimately be arrived at through progressive developments in science. Probably we are at the beginning rather than the end of our understanding of the Universe.

Supernaturalism

The view that living organisms originated on our planet as a result of a supernatural event, seemingly not susceptible to scientific description, is a feature of many religions. Certainly, the origin of the Universe as we know it remains mysterious, and ideas that it developed from a singularity, or even literally from 'nothing', leave basic questions still to be answered, and we must await the further development of cosmological theories in the light of new insights in fundamental physics.

It is quite legitimate to ask whether, in the sort of world we have good reason to believe existed a few thousand million years ago, living things could have arisen without the intervention of a supernatural power, from the continued interaction of different forms of matter and energy. We can attempt to discover whether, given our sort of universe, we can agree with Lucretius that 'Nature, free at once and rid of her haughty lords, is seen to do all things spontaneously of herself, and without the meddling of the gods'. We must recognize, however, that our view of Nature might have to undergo profound changes and refinements in order to accommodate all the phenomena of biology, including consciousness, and we must pay heed to the junction between observational biochemistry and quantum theory referred to by Bernal.

New naturalistic approaches

As recently as 1983, the physicist Paul Davies wrote in his book, *God and the New Physics*,

> How can we weight the credibility of the two explanations for the origin of life (or any other highly ordered system): that life is the product of intelligent, but natural, manipulation by a superbeing, perhaps the supreme being (God), or that life is the end result of mindless self-organizing processes (like the appearance of ordered convection patterns in Jupiter's atmosphere)? Neither explanation is without difficulty.

Fred Hoyle, in his recent writings, has expressed his conviction that a 'chance' origin of life on the Earth is so improbable that for practical purposes it could never have happened. Even if we widen the stage for the origin of life from our Earth to the universe at large, Hoyle (1983) believes, a spontaneous origin of living from non-living matter appears to be extremely improbable. In his recent highly controversial but fascinating book, *The Intelligent Universe*, he discusses the possibility of there being different orders of 'intelligence' in the Universe, even including a directing effect – through individual quantum events – of the future on the past. He suggests that 'The intelligence responsible for the creation of carbon-based life in the cosmic theory is firmly *within* the Universe and is subservient to it. Because the creator of carbon-based life was not all-powerful, there is consequently no paradox in the fact that terrestrial life is far from ideal.'

We shall see later that there are good reasons to believe that arguments based on the supposed extreme improbability of the 'spontaneous' development of organisms may not be soundly based. If it is assumed that atoms and molecules at all times may randomly associate in any possible patterns, then the probability of self-organization proceeding up to the biological level might reasonably be considered to be vanishingly small. However, new developments in the theory of irreversible thermodynamics during the past three decades have cast new light on the behaviour of far-from-equilibrium systems, and appear to have provided a description of processes by which 'order' might be generated from 'chaos'. Consequently, the Universe may appear to act in some respects in an 'intelligent' or 'inventive' manner (see Denbigh, 1975). It should be remembered that our understanding of the way in which the human brain works is at present rudimentary. It is possible that some events in the brain associated with the generation of new ideas and creative thought are basically similar to processes leading to the so-called 'self-organization' which may be observed in chemical systems of a simpler nature, and that there may be similarities in principle between at least some aspects of the inventiveness of organisms and the inventiveness of the Universe at large.

Lithopanspermia, radiopanspermia and directed panspermia

The suggestion has often been made that life may not have originated on the Earth itself, but that 'seeds' or dormant forms of organisms have in some way been distributed in space and have germinated and grown on reaching any suitable planet, such as the Earth.

The adherents of the theory of *lithopanspermia* have suggested that meteorites are the means by which life is transported from one celestial body to another. Numerous attempts have been made to cultivate micro-organisms from meteoric fragments, but results have never been convincingly positive. Indeed, the high probability of contamination of such fragments from terrestrial sources raises great technical and interpretative problems.

The theory gained the support of some eminent scientists in the nineteenth century. The physicist Lord Kelvin, for example, considered it probable that there were countless seed-bearing stones in space, perhaps originating from collisions of life-bearing planets with other bodies. Helmholtz believed that organisms inside meteorites might be protected sufficiently from the harmful heating effects resulting from passage through the atmosphere, the interior remaining relatively cool although the surface of the meteorite was incandescent. The presence of hydrocarbon compounds in meteorites was thought to be due to the former activities of living organisms, but it is now clear that they can be formed by other means including, for example, the reaction of metal carbides with meteoric constituents. The presence of amino-acids and other molecules of types found in organisms is now explained on non-biological grounds. From time to time, claims have been made that meteorites sometimes contain structures morphologically resembling known micro-organisms. In some instances these have been shown to be the result of terrestrial contamination, and there is as yet no convincing evidence for the presence of micro-organisms of extra-terrestrial origin in meteorites. The search will no doubt be continued with more refined techniques (and see Ch. VIII).

The great Swedish scientist Svante Arrhenius (1909), developed the theory of *radiopanspermia*, suggesting that minute germs might be driven from place to place in the Universe by radiation pressure. This is physically possible but it seems likely that an unprotected germ or spore exposed to intense radiations in space would be killed in a fairly short time. Recently, however, new arguments in favour of the presence and survival of organisms in space have again been put forward in the lively and provocative speculations of Fred Hoyle and his collaborator N.C. Wickramasinghe (see Ch. VIII).

Hoyle and Wickramasinghe, after studying the absorption spectra of interstellar material, came to the conclusion that in addition to numerous kinds of organic molecules, whose existence is well established, large molecules such as those of starch and cellulose, and even living bacteria are present in space. This interpretation of the spectra is disputed by other scientists, who claim that absorption by other materials is a more likely explanation. Hoyle has further suggested that comets may be a source of organic molecules and of viruses and bacteria, and has advanced a theory that organisms arriving from space start epidemics of disease (e.g. influenza) in humans. Up to the present, this theory has not received any significant support, and conventional epidemiologists and micro-biologists feel that Hoyle's arguments based on the patterns of diseases in populations are less convincing than the current views of terrestrial origin based on microbial variation and fluctuating population susceptibilities. On the other hand, there are reasons to believe that the synthesis of complex molecules in space from simple starting materials, such as formaldehyde, is a real possibility. Near absolute zero (the coldest temperature possible), some chemical reactions that would not occur at higher temperatures may take place because of quantum mechanical effects, since a phenomenon known as quantum tunnelling can enable the particles concerned to evade the normal activation energy barrier. In effect, the wave-like characteristics of the particle are operative, for when the 'width' of the barrier is smaller than the wavelength of the particle, there is a high probability of tunnelling (see Goldanskii, 1986). In the cold depths of space, the synthesis of the simple starting materials might be initiated by the energy of cosmic rays, and tunnelling could lead to increasing complexity of molecules. Goldanskii points out, however, that tunnelling can also promote destruction of chemical bonds by penetration of activation barriers that are tending to prevent the occurrence of destructive processes. This might reduce the possibility of long-term viability of organisms in the frozen state, if they were not shielded from radiation or other reaction-triggering influences.

The Hoyle and Wickramasinghe theory, which will be considered again later, envisages the production of prebiological and simple biological systems as going on more or less continuously in

Table 1 Some of the molecules in interstellar space

Cyanogen	Dimethyl ether
Hydrogen cyanide	Methanol
Cyanamide	Formic acid
Cyanodiacetylene	Methyl formate
Methylcyanoacetylene	Carbon monoxide
Cyanoethane	Carbon monosulphide
Cyanoethynyl	Ammonia
Formaldehyde	Hydrogen sulphide
Ethanol	Water

Hydrogen cyanide molecules can associate, and then by reacting with water give rise to amino-acids, purines and pyrimidines (including the bases found in nucleic acids). Cyanide can therefore be the sole source of carbon for three important molecular types of fundamental biological importance.

Cyanide and cyanoacetylene can act as starting materials for the synthesis of pyrimidines.

Formaldehyde can under some circumstances, in the presence of clays, give rise to sugars, but the significance of this for biopœsis* has been questioned, as it requires higher concentrations of formaldehyde than were probably available.

If heated in the presence of cyanamide, fatty acids, phosphate and glycerol can react to produce phospholipids, and these can assume the form of vesicles in which other organic molecules can be trapped.

These points are mentioned to indicate a few possibilities, but there is no certainty that these reactions actually played critical roles in prebiotic chemistry. It is, however, apparent that there are known possible mechanisms for the genesis of important starting materials for life, and that the materials involved in the reactions are widely distributed in the Universe.

---

*Pirie (1954) proposed that the process of production of living from non-living matter should be called biopœsis, a term we shall use.

interstellar space, rather than in so-called 'primordial soups' on the surfaces of planets. Interstellar dust clouds are considered to be the regions where the transformations occur. Comets formed from these materials may provide a suitable environment for the further elaboration of primitive organisms, and provide protection from potentially damaging radiations in space. The Earth could have been seeded with organisms during an encounter with a comet, or by dust from comet tails, and such seeding may be continuing. Hoyle would add that the process of development seems to require the intervention of some kind of 'intelligence' of a directive nature, a supposed requirement which, as was mentioned above, might be seen in a different light if we had a more complete understanding of the formation of new dynamic states of matter in far-from-equilibrium conditions. Prigogine (see Prigogine and Stengers, 1984) has called dynamic states of this kind *dissipative structures*. Their formation requires energy, and the energy dissipated in the formation of ordered structures is in effect playing a constructive role. Features able to produce effects in some ways similar to those ascribed by Hoyle to an 'intelligence' might therefore be immanent in the fields of energy and matter in the Universe. This could lead to constructive activity, although the outcome would be unpredictable, and not in conformity to a design, as Hoyle seems to have in mind.

In one place Hoyle (1983) somewhat unflatteringly refers to the 'junkyard mentality' of scientists who favour a terrestrial primordial soup as the site of the origin of organisms. He writes,

> A junkyard contains all the pieces of a Boeing 747, dismembered and in disarray. A whirlwind happens to blow through the yard. What is the chance that after its passage, a fully assembled 747, ready to fly, will be found standing there? So small as to be negligible, even if a tornado were to blow through enough junkyards to fill the whole Universe.

One can certainly agree with Hoyle that the generation of an aircraft in this way is inconceivable, but it must be pointed out that the analogy with the supposed origin of organisms from simpler chemical beginnings is invalid. Complex organisms would not be 'blown together' in any manner resembling what Hoyle had in mind. The formation of quite simple chemical structures would in fact influence and 'direct' what happened next in an environment containing many different starting materials. Reaction rates would be influenced by catalytic effects and a prebiological chemical selection process would operate. The 'tornado' of solar energy would favour the generation of dissipative structures. As Prigogine and Stengers (1984) point out,

31

life no longer appears to oppose the 'normal' laws of physics, struggling against them to avoid its normal fate – its destruction. On the contrary, life seems to express in a specific way the very conditions in which our biosphere is embedded, incorporating the nonlinearities of chemical reactions and the far-from-equilibrium conditions imposed on the biosphere by solar radiation.

These authors would agree, however, that the problem of biological order, involving the transition from molecular activity to supra-molecular order at the cellular level, has not been solved in all its aspects.

Some authors, while believing that the origin of organisms from primordial mixtures of various relatively simple molecules is possible, consider that the Earth was not an ideal site for this process to occur. Nevertheless, it might have occurred on other planets where conditions were more suitable than on the primitive Earth. This has led to the suggestion that the Earth might have been seeded by micro-organisms sent into space by a technologically advanced civilization on another world. This process has been termed *directed panspermia*, and has been proposed by Francis Crick (1982), one of the Nobel laureates associated with the elucidation of the structure of DNA, and Leslie Orgel, a prominent worker in fields related to possible mechanisms of the origin of living organisms. It should be noted that the suggestion was a serious one, and was not put forward facetiously. The age of the Universe is not known with certainty, and the time that has elapsed since the postulated origin in a 'Big Bang' has been a matter of some controversy. Assuming that there was in fact a Big Bang, the age of the Universe is probably greater than 10,000 million years. The Earth is generally considered to be about 4500 million years old. It is possible, therefore, that life might have developed on some other planet formed as long ago as 9000 million years, and that the inhabitants sent off some form of space vehicle containing simple micro-organisms, perhaps bacteria, with the intention of seeking out and seeding other suitable planets. Interestingly, the fossil record suggests that prokaryotic organisms – the simplest organisms, resembling bacteria – appeared early in the Earth's history, more than 3500 million years ago, whereas the structurally more complex eukaryotic organisms (the higher organisms) did not evolve until a little more than 1000 million years ago. This suggests that whatever led to the appearance of the first organisms on Earth took much less time than the evolution of eukaryotes from prokaryotes.

The most that we can say about the theory of directed panspermia at the present time is that it is not disproved by any known facts. A decision between the theory and the conventional ideas of the terrestrial origin of life will have to await further development of our knowledge of the history of the Earth and the prebiological systems that might have existed on the Earth. As Crick remarks, 'The essential difficulty, then, is not so much the nature of the theory, but the extreme paucity of the relevant evidence.'

The theories of lithopanspermia, radiopanspermia and directed panspermia are not strictly relevant to the question of the *origin* of life in the Universe as a whole, but they must be seriously considered as possible explanations of the appearance of life on any one planet. It is possible that if life were to arise independently on different worlds in a solar system, some interchange might occur. This could even lead to the elimination of one form of life in favour of another from an outside source. As our own Solar System, with the very doubtful exception of comets, seems likely to be sterile, apart from the Earth, we shall probably not be able to put any theories of this kind to the test in the near future.

SECTION 2
The constitution of living organisms

Before we can discuss the problem of the origin of living organisms in greater detail, it will be necessary to review briefly the constitution of organisms as we know them on Earth. For brevity, it is convenient to refer to 'life', as for example in 'the origin of life', and it should be noted that when we do so we are not intending to imply that there is an entity 'life' apart from the 'material' organisms with which we are acquainted, and which are studied by biologists and molecular biologists. It was remarked earlier that a full understanding of life will require a more profound knowledge of the nature of atoms and the subatomic world, and probably, it might be added, of the nature of space and time.

Elements in living organisms

The elements that constitute living things are all to be found in inorganic nature – there are no elements which are peculiar to living organisms. The Russian scientist, Vinogradov, identified about sixty elements known to contribute to the formation of organisms, but there are wide variations in the quantities of some elements, not all of the sixty being found in all organisms examined. Carbon, hydrogen, oxygen, nitrogen, phosphorus and sulphur are important constituents of complex molecules present in all organisms, and sodium, magnesium, chlorine, potassium, calcium, iron, manganese, copper and iodine have regularly been found. It will immediately be apparent that organisms are distinguished by the way in which the constituent atoms are arranged to give an *organized* structure, rather than by the nature of the individual constituent atoms (although we must remember that the electronic arrangements of the various atoms make possible the necessary chemical interactions). One of the great problems in the study of the origin of life is to answer the question of how this organization occurred, giving rise not only to the giant molecules found in organisms, but to complex, functioning and ultimately even thinking beings. On the face of it, this seems so unlikely to have occurred without the intervention of some outside organizing 'force' or influence that it is small wonder that many people, including some leading scientists, maintain that our knowledge of the Universe is still too rudimentary to permit a definitive answer.

Life on Earth is strongly dependent on certain features of the carbon atom, and is therefore referred to as *carbon-based life*. Scientists who believe that only carbon-based life is possible, and that other life forms based on different chemical structures are extremely improbable or impossible, have been termed *carbaquists* by Feinberg and Shapiro (1980), a physicist and biochemist who take a broader view of the possibilities of constructing organisms. For the moment, we shall consider carbon-based organisms in more detail, and defer a consideration of other proposed life-forms to a later chapter.

Carbon

Carbon is important because the formation of the crucially significant large molecules we find in living organisms depends on the ability of carbon atoms to form stable chains and ring-structures, to which other elements are attached.

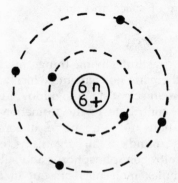

Fig. 3 The carbon atom

The nucleus of the carbon atom contains 6 protons and 6 neutrons, so that 6 electrons are necessary to make up an electrically neutral atom. Two electrons are present in the first shell and 4 in the second (Fig. 3). Since the second shell can accommodate up to 8 electrons, the possibility of interaction with other atoms exists. The result of the interaction of carbon and hydrogen is shown in Fig. 4, and it will be seen that up to four hydrogen atoms can attach to one carbon atom. The four hydrogen atoms contain between them 4 electrons which can be shared with the carbon atom. In the resulting compound, *methane*, CH_4, the carbon atom now has, in effect, 8 electrons in the outermost shell, and each hydrogen 2 electrons, so that the atoms now have their outer shells effectively complete. They have attained what is termed 'noble gas structure'. Each pair of shared electrons constitutes a form of chemical bond between the atoms known as a *covalent bond*. A methane molecule contains only one carbon atom but other related, more complicated, molecules with more than one carbon atom can be formed, for example ethane, C_2H_6, propane, C_3H_8 and so on (Fig. 5). Compounds consisting wholly of carbon and hydrogen are called *hydrocarbons*. Bonds indicated by single lines in the formulæ in Fig. 5 are known as *single bonds*. Other more complex bonds may form between carbon atoms and between carbon and certain other elements, so that in addition to single bonds there may be *double* or *triple* bonds, as in ethylene and acetylene respectively (Fig. 6). Compounds which contain double or triple bonds are termed *unsaturated*, and these bonds are relatively reactive sites in the molecule. Other types of chemical bonding occur but will not be dealt with here. It can be seen that the interaction between atoms to form compounds is not haphazard but is constrained by the electronic structure of the atoms concerned.

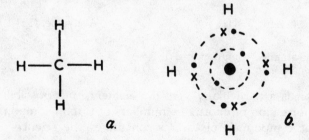

Fig. 4 The structure of methane

a. Conventional structural formula
b. Diagram illustrating electron sharing
x = Electron from hydrogen atom
● = Electron from carbon atom

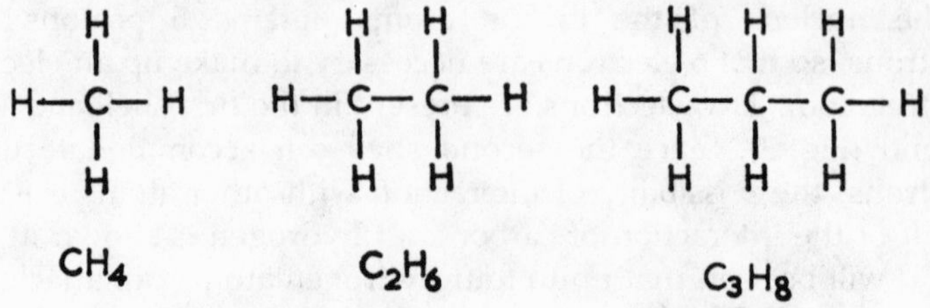

$$H-\underset{\underset{\displaystyle H}{|}}{\overset{\overset{\displaystyle H}{|}}{C}}-H \qquad H-\underset{\underset{\displaystyle H}{|}}{\overset{\overset{\displaystyle H}{|}}{C}}-\underset{\underset{\displaystyle H}{|}}{\overset{\overset{\displaystyle H}{|}}{C}}-H \qquad H-\underset{\underset{\displaystyle H}{|}}{\overset{\overset{\displaystyle H}{|}}{C}}-\underset{\underset{\displaystyle H}{|}}{\overset{\overset{\displaystyle H}{|}}{C}}-\underset{\underset{\displaystyle H}{|}}{\overset{\overset{\displaystyle H}{|}}{C}}-H$$

$$CH_4 \qquad\qquad C_2H_6 \qquad\qquad C_3H_8$$

a. Methane   b. Ethane   c. Propane

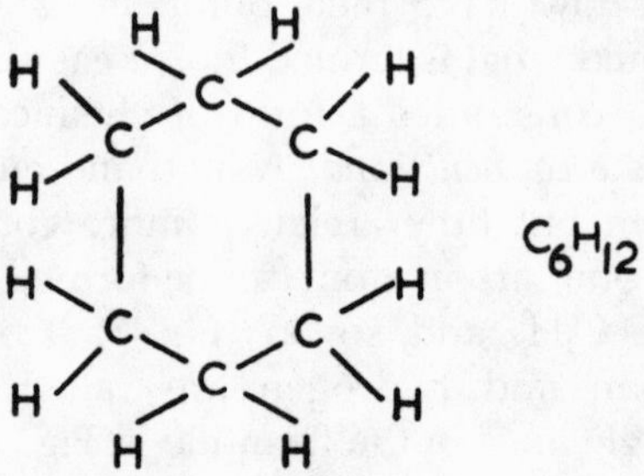

$$C_6H_{12}$$

d. Cyclohexane

Fig. 5 Carbon atoms form chains and rings*

Ethylene, C_2H_4 Acetylene, C_2H_2

Fig. 6 Unsaturated compounds

Carbon is a particularly versatile element, and possibilities exist for the formation of enormous numbers of carbon compounds. The study of compounds of carbon constitutes the science of *organic chemistry*. Carbon atoms linked to each other and to other elements form the 'backbone' of many molecules of biological importance, a few of which we shall consider in more detail.

*The simple molecular formulæ (CH_4, C_2H_6, etc.) show the numbers of atoms per molecule, but do not indicate structure. The structural formulæ show how the atoms are arranged, but because they are two-dimensional, the representation is necessarily limited. Three-dimensional molecular models can give a truer picture.

36

Major molecules of living organisms

Proteins

In this diet-conscious age, most people have heard of proteins, fats and carbohydrates. All are carbon compounds, all occur in living organisms, and any theory of the origin of life and its development into present-day forms must account for their existence, as well as for the occurrence of other large molecules, for example nucleic acids.

Proteins, found in all living things, form a class of chemical compounds with a similar underlying structural pattern, but differing in details of molecular arrangement. The elements of which they are made up are carbon, hydrogen, nitrogen, oxygen and sometimes phosphorus and sulphur. Analysis shows that proteins can be regarded as consisting of simpler units, *amino-acids*, linked together in chains which may be folded or coiled. The shape the molecules assume is partly dependent on a special type of bonding produced by hydrogen atoms. This *hydrogen bonding* is electrostatic, the positive proton of the small hydrogen atom being sufficiently exposed to be able, in some situations, to attract certain other atoms. Hydrogen bonds have only about 10 per cent of the strength of the covalent bonds described earlier but they are of great importance in affecting the shape and possible functions of molecules. They occur in other settings in organisms, for example in nucleic acids, and also are important in the structure of water, a vital solvent for carbon-based life.

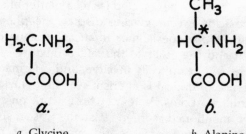

Fig. 7   *a*. Glycine.   *b*. Alanine

The simplest amino-acid is *glycine* (Fig. 7), and a slightly more complicated one is *alanine*. Altogether, about twenty amino-acids are important as 'building blocks' for the proteins of organisms. Examination of the formulæ for glycine and alanine reveals a difference of great significance. In alanine, the carbon atom marked * has four different chemical groups attached to it: $-CH_3$, $-NH_2$, $-H$ and $-COOH$. In glycine there is no carbon atom with four

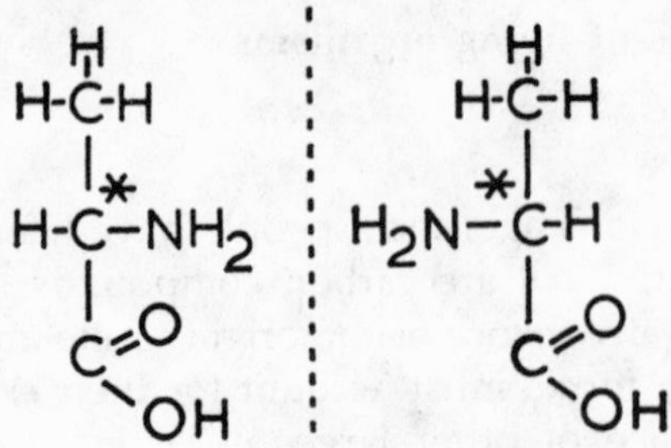

$$
\begin{array}{cc}
\overset{\displaystyle H}{\underset{\displaystyle H\text{-}C\text{-}H}{|}} & \overset{\displaystyle H}{\underset{\displaystyle H\text{-}C\text{-}H}{|}} \\
\underset{\displaystyle H\text{-}\overset{*}{C}\text{-}NH_2}{|} & \underset{\displaystyle H_2N\text{-}\overset{*}{C}\text{-}H}{|} \\
\underset{\displaystyle \overset{O}{\underset{OH}{C}}}{} & \underset{\displaystyle \overset{O}{\underset{OH}{C}}}{}
\end{array}
$$

Fig. 8 'Mirror-image' molecules

different groups attached. A carbon atom of the type C* is called an *asymmetric* carbon atom, and it is possible to obtain two distinct forms of a structure containing an asymmetric carbon atom, related to each other as mirror images. In two dimensions, the molecular forms can be represented as in Fig. 8. The nomenclature L(+) and D(−) has been used to distinguish these forms, and for the present it is sufficient to note that the (+) and (−) indicate that solutions of the compounds will rotate the plane of polarized light to the right and left respectively. If a molecule contains more than one asymmetric carbon atom, there will be two possible forms at each site. The existence of more than one molecular form, made up of the same number of atoms, is an example of chemical *isomerism*, and since the particular molecules we have considered influence the behaviour of polarized light, they are called *optical isomers*. One might expect to find both isomers equally represented, but the amino-acids found in living organisms belong to the L series. How this isomeric form came to be selected is still a matter of speculation.

All the amino-acids except glycine display the phenomenon of optical isomerism. Since one isomer rotates the plane of polarized light to the right and the other to the left, a mixture of the two in equal proportions will be optically inactive, and is termed the *racemic* form. The existence of optical isomerism was first elucidated in the nineteenth century by Louis Pasteur, whose work on spontaneous generation was mentioned earlier. Pasteur was responsible for fundamental advances in chemistry and microbiology, and laid the foundations for the medical applications of bacteriology and virology.

In proteins, the amino-acid building blocks are linked by the formation of *peptide bonds* (Fig. 9). Proteins have high molecular weights, and contain large numbers of amino-acid units in each molecule. The sequences of different amino-acids in a protein molecule determine its shape and functions, and small alterations, even sometimes of a single amino-acid, may render a protein

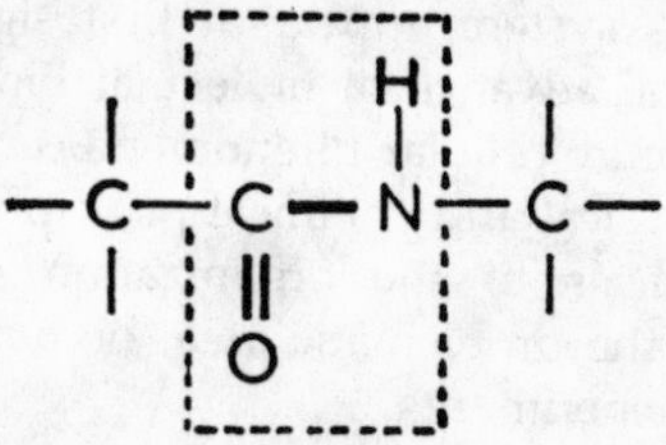

Fig. 9 The linkage within the dotted rectangle is a 'peptide bond'

biologically inactive or inefficient. In the course of evolution, some aspects of protein structure have been conserved over vast periods, whereas others have changed. Environmental factors, including temperature, acidity and presence of reactive atoms or molecules may strongly affect proteins.

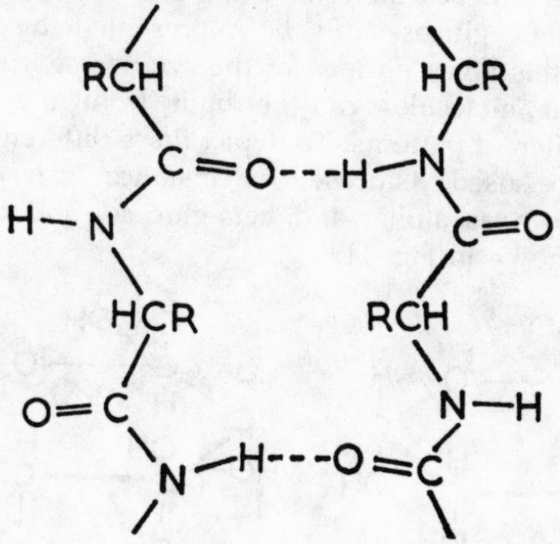

Fig. 10 Part of a protein molecule. R = chemical groups not shown in full. − − − between H and O atoms = hydrogen bonds

In living organisms, proteins are involved in both structure and function. Of great importance is the ability of certain proteins (*enzymes*) to speed the rates of chemical reactions, often with a high degree of specificity. Complex reaction chains in cells may involve many different enzymes acting in a coordinated manner, and it is only in recent decades that a beginning has been made in understanding the ways in which the complicated chemical changes going on in organisms are controlled and *organized*. There are *ordered* structures, such as crystals, which do not show *organization* in the sense in which we find it in metabolizing organisms. Even at the level of single-celled organisms, the delicate controls necessary to maintain the living system appear at first sight almost incredibly complicated, but the advance of molecular biology has shown that many formerly obscure cellular phenomena can be explained on the basis of known physical and chemical principles. The passage from mixtures of chemicals to the organization of the first cells or protocells remains the problematic area, because of the difficulty of devising relevant experiments.

Carbohydrates

In various forms, carbohydrates are used in daily life. Cane sugar and glucose are fairly simple carbohydrates, and starch is a more complicated one, a *polysaccharide*, this term implying that starch molecules are built up from many sugar, or *saccharide* units. Carbohydrates consist of carbon, hydrogen and oxygen atoms joined to form molecules, the ratio of the number of hydrogen atoms to oxygen atoms in each molecule being 2:1.

A molecule of glucose can be represented by the formula $C_6H_{12}O_6$, but this gives no idea of the arrangement of the atoms. Different sugar molecules can be built from the same atoms arranged in different patterns. To depict these differences, *structural formulæ* can be used. Glucose, for instance, can exist as ring structures known as alpha- and beta-glucose, and the structural formulæ are shown in Fig. 11.

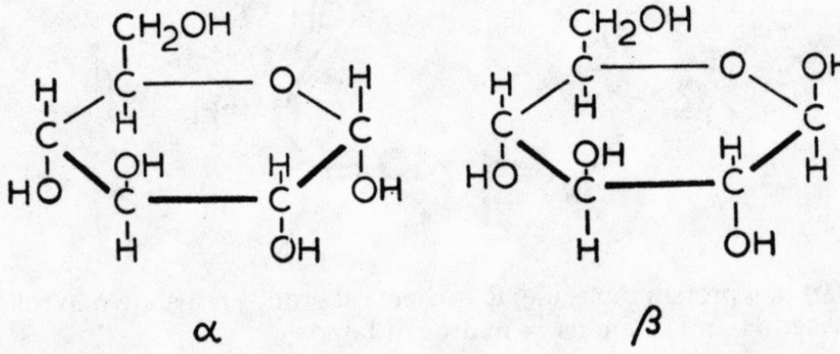

Fig. 11 Structural formulæ of α- and β-glucose

The chemistry of carbohydrates is complicated, and many subtle variations of molecular structure are possible. Large polysaccharide molecules consist of many, even hundreds, of sugar units linked together. The wide range of carbohydrates produced by plants and animals is an indication of their importance for present-day organisms.

Fats

These form another class of compounds built up from carbon, hydrogen and oxygen atoms and, as with other compounds, the properties of fats are determined by the arrangement of the constituent atoms. The structural formula of a typical fat is shown in Fig. 12. Fats containing double bonds between some of the carbon atoms are termed *unsaturated*.

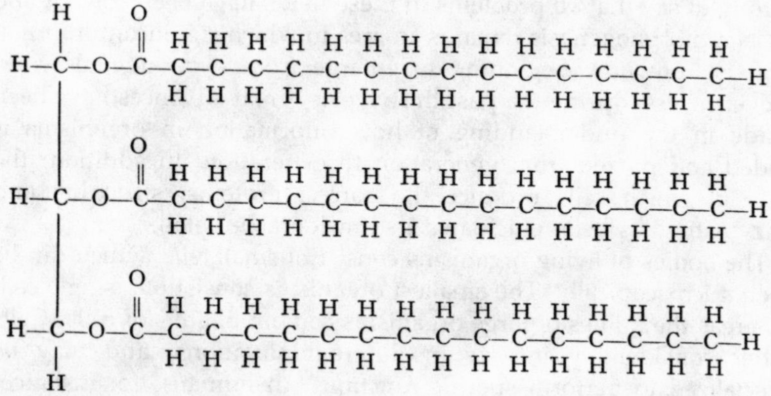

Fig. 12 A fat molecule

Fats are *esters* of the trihydric alcohol *glycerol* ($CH_2OH.CHOH.CH_2OH$) with saturated or unsaturated *fatty acids*. Fatty acids consist of chains of carbon atoms with attached hydrogen atoms, with at one end a $-COOH$ (carboxyl) group. They provide the long chains of carbon atoms shown in Fig. 12. The class of fatty substances, or *lipids*, includes a variety of compounds, fatty acids, triglycerides, glycolypids and phospholipids, that are either insoluble in water or poorly soluble, but they do dissolve in certain other solvents. The cells of living organisms are bounded by membranes in which both lipids and proteins play important rôles. The properties of lipid molecules, including their capacity to aggregate rather than to disperse in water, are of fundamental importance in the production of membranes able to form boundary structures around and within cells. In the production of the first cellular organisms, compartmentalization of components was undoubtedly a necessary feature, and the properties of lipids would permit them to play a major rôle in this process.

Nucleic acids

Typically, living things grow and from time to time multiply. In order to grow, they must be able to utilize foods as a source of material for the construction of the molecules they need for their cellular components, and these components must be arranged into the appropriate patterns. When organisms multiply, new individuals closely resembling the parent or parents are produced and this requires some method of duplication of structure, or of passing on *information* on how growth is to proceed. Here, there are problems of production of the correct materials, structural molecules, and of the appropriate form, the process of *morphogenesis*. It would be wrong to say that all problems in these fields have been solved, and indeed morphogenesis remains an area in which significant information is only now beginning to emerge from experimental work. Nevertheless, during the past thirty years, great advances have been made in the understanding of how information in organisms is coded and passed from generation to generation. In addition, the way in which cells produce the correct proteins, structural and enzymatic, has been elucidated in considerable detail.

The bodies of living organisms consist of small *cells*, which can be seen microscopically. The smallest organisms consist of a single cell, whereas the bodies of large organisms contain billions of cells. Cells differ considerably in size in different situations, and may be specialized to perform specific functions. In humans, for instance, we find muscle cells, brain cells, blood cells, liver cells and so on. The diameters of cells are commonly in the range of $1/100$ to $1/20$ mm, but there are wide differences in shape, some being of great length in proportion to their width. Microscopically, in higher organisms (*eukaryotes*), each cell (Fig. 13) is seen to consist of a dense central region, or *nucleus*, which is membrane bounded, surrounded by an outer zone, the *cytoplasm*, bounded by the *cytoplasmic membrane*. The cytoplasm contains various *organelles*, the *mitochondria* and *ribosomes*, active respectively in deriving energy from nutrients and building up new protein molecules. In the 'simplest' cells (*prokaryotes*) there is no membrane around the nuclear material and mitochondria are absent, but the cytoplasm is bounded by a cytoplasmic membrane. Indeed, the whole prokaryotic cell has been likened to a mitochondrion. Some biologists believe that the mitochondria of eukaryotes have been developed from prokaryotes associating with other cells (Margulis, 1971, 1981, 1985). In the cells of green plants, we also find *chloroplasts*, complex organelles responsible for the process of *photosynthesis*.

Two kinds of nucleic acid have been identified in cells, (1) *deoxyribonucleic acid* (DNA), containing as a component of its molecule *deoxyribose*; and (2) *ribonucleic acid* (RNA), containing *ribose*. One of the most exciting developments in biology during the past thirty years has been the discovery and increased understanding of the role of nucleic acids in heredity and the control of cellular functions.

The DNA, which is a major component of cell nuclei, is built in such a way that it can act as a store of coded information for the functioning and reproduction of cells. The structure of DNA was

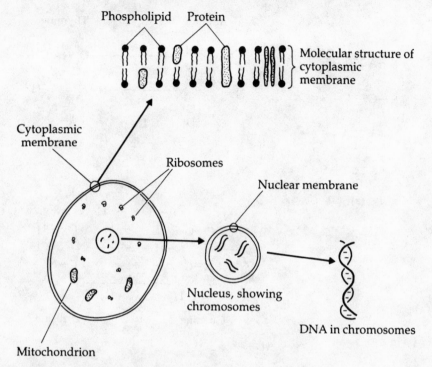

Fig. 13 Aspects of eukaryotic cell structure

finally established by the work of Watson and Crick, and the story of the 'double helix' is well known (Watson, 1968). Like proteins, nucleic acids have giant molecules which can be analysed into simpler units. The elements in nucleic-acid molecules are carbon, hydrogen, oxygen, nitrogen and phosphorus. The DNA molecule has a 'backbone' of sugar units (deoxyribose) linked by phosphate groups, and attached to this backbone are the *bases*, adenine, cytosine, guanine and thymine (Fig. 14). A unit of structure consisting of base-sugar-phosphate is called a *nucleotide*, and nucleic acids can be regarded as chains of linked nucleotides. The molecules of DNA are helical, consisting of two DNA chains wound around a common axis and linked by hydrogen bonding between the protruding bases (Fig. 15). In these linkages, adenine always links with thymine (A–T), and guanine with cytosine (G–C). The members of such base pairs and the corresponding nucleotides and strands are said to be *complementary*. This complementarity is dictated by the structure of the bases and the consequent hydrogen bonding (Fig. 14).

When a cell divides, there must be duplication of the various structures. Separation of the two strands of the DNA with exposure of the bases allows the building up of new strands by association

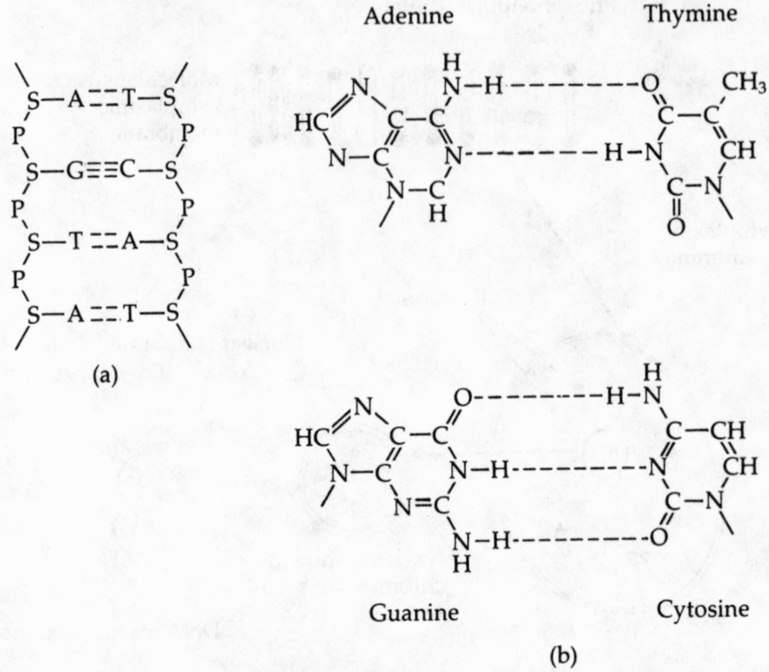

Fig. 14 Aspects of DNA structure

(a) Shows 'backbones' of DNA molecule, sugar (S=deoxyribose) units linked by phosphate (P) groups.
(b) Shows formation of hydrogen bonds between complementary bases. Note that there are three hydrogen bonds between G and C and two between A and T.

with complementary nucleotides. The two halves of the original double helix act as patterns for the building of new strands, and if no errors are made, ultimately two·identical double spirals will be produced. This process requires the participation of numerous enzymes, which are proteins.

The sequence of the four types of nucleotide with their respective bases (G, C, A or T) has been found to be, in effect, a series of 'words', each of which is three nucleotides ('letters') long and specifies a particular amino-acid, the order of the 'words' along the DNA strand corresponding to the sequence of amino-acids in a particular protein. A section of a DNA strand specifying a complete protein molecule is called a *gene*. However, the DNA code is not translated directly into protein structure; instead, the synthesis of proteins is achieved through intervention of RNA molecules.

Molecules of RNA differ in important respects from those of DNA. The sugar in the backbone is ribose, and in place of the base thymine, another, called *uracil* is substituted. As with thymine, a nucleotide containing uracil is complementary to one containing

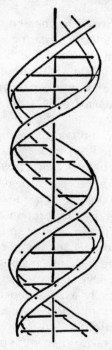

Fig. 15 Helical form of the DNA molecule. The hydrogen bonding between complementary base pairs is indicated by the horizontal bars. For details see Fig. 14

adenine. Moreover, RNA is usually single stranded, and the molecules are much shorter than those of DNA. When a gene is to be used for protein production, the two strands of DNA are separated and a molecule of RNA is built along the strand bearing the code for the protein, each base of the RNA being complementary to that of the corresponding DNA nucleotide. This process is called *transcription*, as the DNA code is transcribed into RNA code. The RNA so produced is called *messenger* RNA (mRNA), as it carries the message from the DNA gene to the site where protein will be produced. The message is in the form of triplets of nucleotides (e.g. UUU, UCC, GGA) each called a *codon* and specifying a particular amino-acid. The messenger moves into the cytoplasm, and there encounters RNA molecules of another class, *transfer RNA* (tRNA), whose molecules are twisted into looped forms as a result of some base pairing. At the looped end of a tRNA molecule is a triplet of unpaired bases, termed the *anticodon*, rendering it able to bind with a complementary codon on a messenger RNA strand. The other end of the tRNA molecule is able to bind with the appropriate amino-acid indicated by the anticodon, again with the assistance of an enzyme. The tRNA bearing the amino-acid attaches to the comple-

45

mentary codon. In this way, the messenger RNA will have strung along its length tRNA molecules bearing amino-acids in the correct sequence for a particular protein. The final joining of the amino-acids to form the protein is brought about by a ribosome, which is an organelle consisting of a number of proteins associated with a third kind of RNA, *ribosomal RNA* (rRNA). The messenger bears 'start' and 'stop' signals to 'tell' the ribosome where to begin and end. As the amino-acid chain is produced, it folds itself into its final form. The 'reading' of the RNA message to permit the generation of the protein amino-acid chain is called *translation*.

It should be noted that recent work has shown that RNA may have catalytic functions of a type formerly considered to be manifested in organisms only by protein enzymes.

The complexity of the process of protein synthesis, and the control necessary to assure its successful conclusion, are staggering at first sight. If we picture the Universe in old-fashioned mechanistic terms, we can hardly believe that anything so extraordinary would arise by 'chance', especially as it is much more complicated than described above, as many protein enzymes are necessary to catalyse various reactions in the process. But, as we have seen, atoms are not little hard balls jostling each other without rhyme or reason. Their structure is itself of great subtlety and dictates certain modes of behaviour. It is true that at the molecular level in a primordial soup the possibilities of random interactions would be enormous, but the whole notion of what might arise by 'chance' has taken on new aspects in the light of recent developments in irreversible thermo-dynamics, and the study of systems far from equilibrium behaving in a non-linear fashion. There is a growing suspicion that what we think of as 'chance' may be the sole generator of novelty, and that the Universe has built-in characteristics favouring the production of patterned processes from seemingly chaotic beginnings (see Prigogine and Stengers, 1984; Crutchfield *et al.*, 1986).

Energy of activation and chemical reactions. Enzymes

Atoms or molecules in a mixture are only able to react if they reach a certain energy level, when they are then said to be in the *activated state*. The energy necessary for the activation of the reactants may be obtained from various sources, for instance heat or ultra-violet radiation.

It is frequently found that the energy of activation for a reaction between two substances is reduced if a third substance is introduced into the reaction mixture, although the third substance does not undergo permanent change in the course of the reaction. This is the phenomenon of *catalysis*, and the substance which leads to

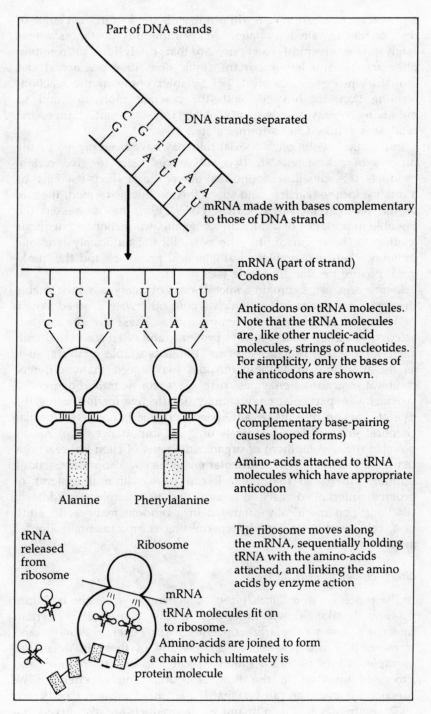

Fig. 16 Steps in protein synthesis

acceleration of reaction rate without itself being lost in the course of the reaction is called a *catalyst*. Many different substances act as catalysts for chemical reactions, so that catalysis is a common phenomenon and has important applications in experimental and industrial processes. Catalysts act by intervening in the reaction, forming transient linkages with the reactant molecules and so modifying energy requirements. After this transient change, the catalyst is restored to its former state.

In chemical evolution, a special rôle may have been played by the process of *autocatalysis*. If two substances react to give certain products, it is sometimes found that one of the products itself acts as a catalyst for the reaction, and so, as this product is formed, the rate of the reaction increases. Calvin (1959, 1969) has discussed the possible importance of autocatalysis for the production of chemicals leading to the origin of life. The possibility of autocatalysis is one difference between prebiological chemical processes and the 'junk-yard' picture painted by Hoyle (see p. 31).

Living organisms contain a special class of catalysts which enable them to carry out reactions which would otherwise proceed slowly or hardly at all at ordinary temperatures. These are the *enzymes* referred to earlier, almost always proteins, and often associated with smaller molecules called *co-enzymes*. Whereas simple catalysts, such as the inorganic manganese dioxide, may speed many different chemical reactions, enzymes display a considerable degree of specificity. A particular enzyme may be effective for increasing the rate of reaction of only certain specific chemicals or closely similar chemical groups. This feature is of great importance, as it makes possible the development of organized chains of chemical reactions that go on in living cells. Particular enzymes speed certain reactions but not others, so that the cell can have within it systems of ordered, interlinked reaction chains, and not merely a muddle of reactions non-specifically catalysed in a random manner. In addition, cell structure is complex, permitting compartmentalization of systems within cells.

Other 'biological' molecules

In the processes of cell *metabolism* – a general term for the building-up and breaking-down processes in the living cell – certain molecules are particularly concerned with energy storage and transfer. It is not possible to discuss any of these in detail but examples will be mentioned to give some idea of the structures and processes involved in the flow and utilization of energy. The functioning organism can be regarded as an *open system* (see Perret, 1952). Nutrients are taken in and waste products are discharged; the inflow of food supplies the materials and energy for life, but the food has to be altered and used for building new parts. The organism is at the same time wearing out and reconstituting itself. It is an open system because it continually exchanges material and energy with its surroundings, and death is marked by cessation of this process of interchange. Special molecules play a key role in the energy transformations of the living process.

Adenosine triphosphate (ATP)

During metabolism, energy is stored for a time in special 'high energy' chemical bonds which are formed in certain phosphorus-containing organic molecules. ATP is an excellent example. The molecule may be represented as in Fig. 17, and the bonds indicated by the sign ~ are high-energy bonds. ATP takes part in various reactions in which phosphate groups are transferred to other molecules, and these transfers are accompanied by release of energy which is used by the cell. In the course of metabolism, ATP is repeatedly handing on phosphate groups to other molecules and then being re-formed by acquisition of phosphate groups from other sources.

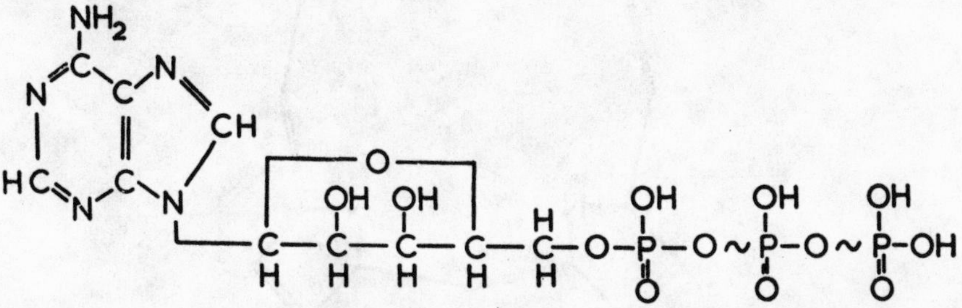

Fig. 17 Adenosine triphosphate

It is possible that simpler phosphate-containing compounds capable of analogous exchange mechanisms were developed at an early stage in chemical evolution, and it has been suggested by Lipmann that carbamyl phosphate, $OC.NH_2.OPO_3$, might be regarded as a first step in biochemical evolution.

Porphyrins

The members of this group of chemically related substances are responsible for important processes in cells. The molecules are variations on the basic structure *porphin* (Fig. 18). There are several different porphyrins, and in their functional forms they are usually combined with a metal and a protein to form highly active compounds. An iron-porphyrin combination is part of the hæmoglobin of blood, which is responsible for the transport of oxygen throughout the body. Many types of cell contain pigments called *cytochromes*, which form part of electron transport systems concerned with cellular respiration, and these are metal-porphyrin-protein compounds. Another iron-porphyrin, the enzyme *catalase*, which is widely distributed in living organisms, breaks down hydrogen peroxide to water and oxygen, the iron facilitating the interaction of the enzyme with hydrogen peroxide. Ions of inorganic iron have some activity against hydrogen peroxide, but one milligram of iron combined in catalase has an activity equivalent to that of about ten tons of inorganic iron. Calvin has used this as an illustration of his view that the fundamental character of the enzymes of present-day organisms results from the evolutionary development of rudimentary powers of catalysis of simple ions or molecules of the prebiological environment.

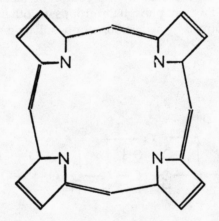

Fig. 18 'Skeleton' of the porphin molecule. The molecule is not shown in full, but the lines, indicating chemical bonds, give an idea of its form. There is a carbon atom at each angle, except where nitrogen is shown.

Chlorophyll, the colouring green pigment of plants, is a magnesium-porphyrin compound (Fig. 19) which helps to trap the energy of sunlight and so make it available to plants for driving metabolic processes. Calvin has pointed out that iron would be unsuitable for this purpose. After chlorophyll has captured a quantum of solar energy, a long-lived excitation is necessary for its functioning, but the magnetic nature of the iron atom would preclude this. Magnesium can have a long-lived excitation, and so mediate energy migration and conversion with high efficiency.

Porphyrins are found as numerous other pigments in nature; for

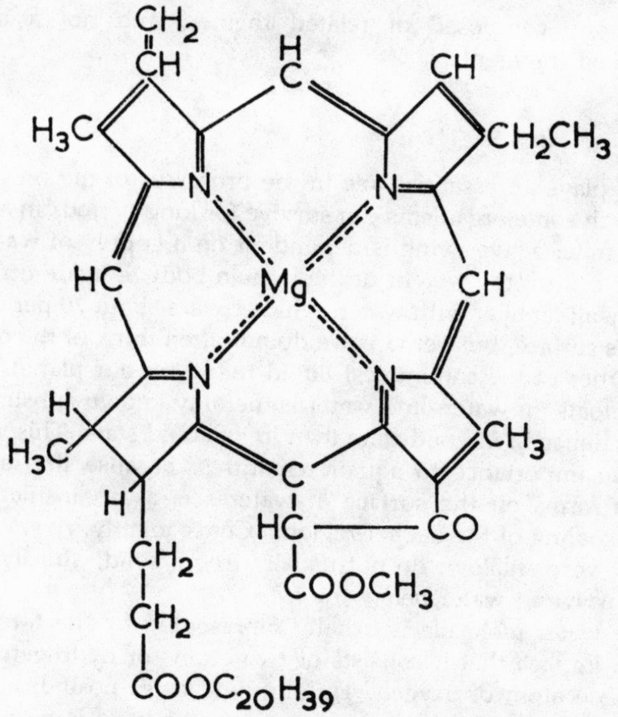

Fig. 19 Molecule of chlorophyll *a*, of plants

example, the copper-porphyrin *turacin* in the feathers of the Turaco bird, and copper-containing *haemocyanins*, which act as oxygen carriers in certain invertebrates. We can see how subtle variations on a chemical theme can lead to the production of a series of fundamentally related but functionally different compounds. We can appreciate, too, that the functional capacities of these compounds are an expression of their structure. In the course of evolution, the most efficient molecules available have, by natural selection at the molecular level, become predominant in biological processes.

In addition to the few classes already mentioned, many other molecules of different kinds are widely distributed in living things. Enzymes containing vitamins of the B group, for example, are found practically universally in cells. Bernal remarked that there is 'an extreme chemical and structural conservatism or inertia of once-established structure or activities. At and above the level of the protozoa there is no essential change in structure or metabolism of the nucleated cell.'

All the variety of living things as we know them on Earth share many features in common at the fundamental level of cell metabolism, and they share a similar genetic code. Life is a symphony composed of related themes, and not a medley of unrelated fragments.

Water and life

Water plays an essential rôle in the processes of life on Earth, and although some organisms can survive for long periods in a dormant, dried state, active living is dependent on a supply of water. About 60 per cent of the weight of the human body is made up by water. We are all familiar with water, which covers about 70 per cent of the Earth's surface, but perhaps we do not often think of the remarkable properties of the commonest liquid found on our planet.

Ice floats on water, and water is the only known substance which is less dense in its solid state than in its liquid state. This property is of vital importance to aquatic creatures, because the layer of ice which forms on the surface of water acts as an insulator against rapid cooling of the deeper regions. Consequently, rivers and lakes, unless very shallow, do not usually freeze solid, and living things can survive in water below the ice.

The water molecule is usually represented by the formula H_2O, which implies that it consists of two atoms of hydrogen combined with one atom of oxygen. This formula gives no indication of the nature of water in the mass, for when we have large numbers of water molecules together, hydrogen bonding between hydrogen and oxygen atoms occurs. In liquid water, hydrogen bonds are being formed, broken and re-formed as the molecules move because of thermal agitation. The hydrogen atom, it will be remembered, consists of a proton with one orbiting electron, whereas oxygen has 8 electrons arranged around the nucleus in two shells, an inner one with 2 electrons and an outer with 6. The electrons of two hydrogen atoms can 'fit' into the outer shell of the oxygen atom and so lead to the formation of a water molecule. Each hydrogen atom is in addition attracted towards the oxygen of another water molecule by hydrogen bonding, so that water can be described as a 'united association' rather than as a mixture of molecules (Fig. 20). As water is cooled, thermal agitation of the molecules decreases, and in due course ice will form. Linus Pauling has proposed that in ice all the water molecules form all possible hydrogen bonds, resulting in a structure in which the component molecules are more widely spaced and so occupy a larger volume than in the liquid. Melting of ice is at first accompanied by breakdown of the structure and some degree of moving together of the molecules, so that the density increases, this process continuing until the temperature has been raised to 4°C, beyond which thermal agitation begins to cause expansion of the liquid. At the boiling point, the rapidly moving molecules fly off in the form of gaseous steam. The boiling point of water is higher than it would be if hydrogen bonding did not occur.

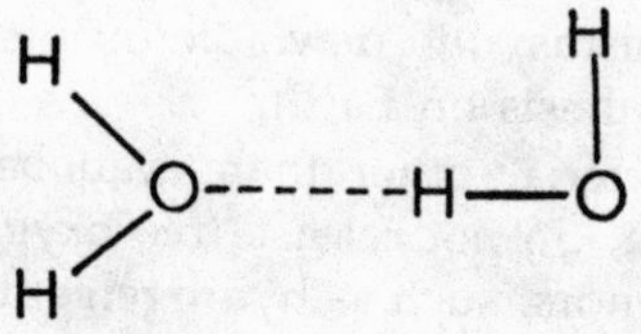

Fig. 20 Water molecules associating. $----$ = hydrogen bond

Water is an excellent solvent for many substances, so the preparation of water of high purity is difficult. Many substances, when dissolved in water, dissociate into positively and negatively charged ions. Common salt – i.e. sodium chloride, NaCl – is present as positively charged sodium ions (Na^+) and negatively charged chloride ions (Cl^-). Substances which dissociate in this way are called *electrolytes*, and water containing them is a good conductor of electricity, unlike pure water.

Water is a major component of living things. In addition to forming a medium for chemical reactions, it takes part in reactions itself and has a profound effect on the detailed organization of living matter. Horne (1971) has argued that the unique ability of water to form polymorphic three-dimensional hydrogen-bonded aggregates and solvation envelopes is a *necessary* condition for the 'realization of the levels of order in form and complexity in function required by carbonaceous biotic systems'. The existence of non-aqueous carbon-based life therefore seems to be unlikely.

Photosynthesis

For the Earth, the Sun is the ultimate source of energy for most living organisms. The trapping of solar energy is achieved by green plants and by some other organisms which contain special chemical systems and structures able to perform this feat. The energy of sunlight trapped by chlorophyll is directed into the synthesis of ATP and is used for the production of carbohydrates from carbon dioxide. In plants, the chemical machinery for the complex reactions involved is located in membrane-bounded intracellular organelles called *chloroplasts*.

Green plant photosynthesis has as one of its effects the splitting of the water molecule, which makes hydrogen available to react with CO_2 to form carbohydrates, and releases oxygen. Carbon dioxide is chemically *reduced* by combining with hydrogen. The oxygen, derived from water, appears as the free gas. It is likely that oxygen released in the course of photosynthesis of this type (also carried out by a group of prokaryotic organisms called *blue-green algae*, or more properly *cyanobacteria*), accounts for the abundance of oxygen in our atmosphere. Marine plants are a major component of the totality of photosynthetic organisms, and may account for as much as 80 or 90 per cent of photosynthesis on Earth.

Photosynthetic bacteria, other than cyanobacteria and a group called *prochlorobacteria*, do not release free oxygen, as their systems use other electron donors, such as hydrogen sulphide (H_2S), but not water (H_2O), in photosynthesis. They mostly contain chlorophyll, differing in details of structure from that of plants. Bacteria included in the group *Halobacterium* utilize not chlorophyll but *Bacteriorhodopsin* as a photosynthetic pigment, and require organic carbon sources. We mention these points to indicate that light can be used by a variety of organisms to promote syntheses in systems that are not identical. If organisms exist elsewhere in the Universe, it is quite likely that some are photosynthetic, acting as primary energy transducers for the biosphere.

Calvin has suggested that photosynthesis involving chlorophyll might have occurred very early in the development of life on Earth, and many workers believe that photosynthesis of the type found in bacteria antedated the type found in plants by a considerable period. Photosynthetic bacteria (*anoxygenic*, not releasing oxygen) are believed to have existed as long as 2.8–3 billion years ago, and cyanobacteria, able to release oxygen, date back more than 2 billion years (here we use billion to mean 1000 million). The amount of free oxygen in the atmosphere increased as a result of oxygenic photosynthesis. Oxygen is quite reactive, and would combine with surface materials on the Earth and so be bound if it were not continuously released back into the atmosphere by some mechanism. Although the break-up of water by ultra-violet light (*UV-photolysis* of water) in the atmosphere produces some free oxygen, it could not account for the quantity needed to bring about the changes over geological periods.

The existence and persistence of large quantities of free oxygen in a planetary atmosphere is an indication that the atmosphere is not in equilibrium but is inherently unstable, and might reasonably be taken as evidence in favour of the existence of photosynthetic organisms on the planet. On the other hand, the absence of oxygen would not rule out the existence of photosynthetic organisms, for not all types release free oxygen. The activities of large numbers of organisms on a planet would be expected to affect the composition of the atmosphere in some way so that it would not be in a state of equilibrium. Lovelock (1979, 1986) has stressed the value of observation of the constitution of planetary atmospheres in searches for extraterrestrial life. He wrote (1986):

When viewed from space in infra-red light, our planet is a strange and wonderful anomaly compared with Mars and Venus. On Earth, the gases of the air are an unstable mixture of hydrocarbons and oxygen . . . we know that the Earth's unstable atmosphere persists. . . . It is this persistent instability that suggests that the planet is alive. . . . The Earth has remained a comfortable place for living organisms for the whole 3.5 billion years since life began, despite a 25% increase in the output of heat from the Sun . . . living organisms have always, and actively, kept their planet fit for life.

Entropy and life

Entropy may be described as a measure of the disorderliness of a system. The second law of thermodynamics states that, in a *closed* system, and so in the Universe as a whole, entropy always increases. Entropy has been used as an indicator of the 'arrow of time'. If we know that at time t_1 the entropy of a system was less than at time t_2, then t_2 was later in time than t_1. Eddington (1935) emphasized that when entropy is used as a 'signpost' for time, it is necessary to take care that a properly isolated system is being considered, for a system can gain order by 'draining it from contiguous systems'.

When organisms grow and develop, they apparently gain organization, and it has sometimes been suggested that they defy the second law of thermodynamics and so lie outside the realm of physics. However, animals gain organization only when they are able to take in food and use the energy stored in it, and plants can use the energy of sunlight. An organism does not remain active if deprived of energy supplies. As Eddington wrote, if we cut off a man's food, drink and air, 'he will ere long come to a state which everyone would recognize as a state of extreme "disorganization" '. In the words of Schrödinger (1944), organisms feed on 'negative entropy', and if we trace back the energy chain from animals to plants, we find that plants use solar energy and it is the organized solar radiations which provide the necessary energy, the 'negative entropy', for the building-up processes of organisms. In isolation, organisms obey the second law of thermodynamics, and it does not seem necessary to postulate 'vital forces' of a non-physical kind to account for the local entropy decrease in the functioning organism. If we consider the organism plus its environment, there is a net increase of entropy during growth and development.

Can life be defined?

Many attempts have been made to frame a brief and satisfactory definition of life, and all seem to have shortcomings. It seems that no matter what list of properties we draw up as characteristic of living things, exceptions are always found. Movement is not peculiar to living things, nor characteristic of all organisms, and neither is growth, for crystals grow. There is general agreement that some things – active animals, growing plants, growing bacteria – are 'alive', but when we consider viruses, which replicate only within living cells, there is plenty of room for argument. Pirie (1937), in an article entitled 'The Meaninglessness of the Terms "Life" and "Living" ', argued that until a valid definition has been framed it is prudent to avoid the use of the word 'life' in any discussion about borderline systems, and to refrain from saying that certain observations on a system have proved that it is or is not 'alive'.

Bernal (1951) some years ago tentatively suggested that, if we limit ourselves to a consideration of the 'life' accessible to our observation on Earth, 'we can for the moment find one common characteristic, the presence of protein molecules, and . . . one common physico-chemical process, the stepwise catalysis of organic compounds carried out practically isothermally by quantum jumps of between 3 and 16 kilo-calories, small compared with the usual jumps of 300 in laboratory chemistry'. More recently, (1967) he proposed a more general description of living processes: 'Life is a partial, continuous, progressive, multiform and conditionally interactive, self-realization of the potentialities of atomic electron states.' Each of these statements emphasizes important aspects of 'life'. Others would place more emphasis on nucleic acids and their rôle in cells. Maynard-Smith (1986) has argued that 'the ability of like to beget like is the most fundamental characteristic of life'.

Horowitz (1957) proposed that living things are characterized by three properties, ability to duplicate, to influence their environment in a way which ensures a supply of materials necessary for their perpetuation, and an ability to mutate randomly and reproduce in the new form. On this view, certain types of complex molecules might be considered to be alive.

It was mentioned earlier that some workers now emphasize the living nature of the whole *biosphere*, or even the whole *planet*, and regard it as in effect one self-regulating living entity, as outlined in the *Gaia* hypothesis put forward by Lovelock and Margulis (see Lovelock, 1979, 1986). Based on this approach, Feinberg and Shapiro (1980) have proposed a definition of a quite general nature: '*Life is fundamentally the activity of a biosphere. A biosphere is a highly ordered system of matter and energy characterized by complex cycles that maintain* or gradually increase the order of the system through an exchange of energy with the environment.' This leaves open the chemical nature of the components of a 'living' system, and would apply to non-carbon based life, if it exists.

Enough has been said to illustrate approaches to framing a definition. Obviously, a confirmed carbaquist might frame a definition in terms that would be too restrictive for someone who took a much more general view of the possibility of finding alternative life chemistries. Fortunately, as Pauling (1957) has remarked, 'it is sometimes easier to study a subject than to define it'.

SECTION 3
Pathways to life

Living things are dynamic systems in which the parts interact in a fairly precise manner. They have self-regulating features but their proper functioning depends on the existence of a suitable range of environmental conditions. The surroundings must not be too hot or too cold, and the correct nutrients must be available. In the course of ages, many different varieties of organism have evolved, and some can tolerate, and may even require, conditions which would be fatal to others.

From a more fundamental viewpoint, the existence of a Universe so constituted physically that it can produce and maintain living organisms has struck some physicists as remarkable, because small alterations in the fundamental constants of nature would have rendered the Universe unsuitable for life as we know it. Paul Davies (1982) has discussed this and related problems in his book *The Accidental Universe*, and there is no space here to consider this matter in detail. What has been called the *weak anthropic principle* is a statement to the effect that *what we can expect to observe must be restricted by the conditions necessary for our presence as observers*. Some scientists support a more extreme view, termed the *strong anthropic principle*, which states that *the Universe must be such as to permit the creation of observers in it at some stage*. In Wheeler's 'participatory Universe', the act of 'observation' (an irreversible amplification) is considered to have a creative role. The rate of expansion of the Universe is important. Had it been at the outset lower, the Universe would have collapsed too soon to permit star formation, whereas had it been higher, the rate of expansion would have been too great to permit gravitational condensation of atoms to form stars. A different gravitational mass for the proton might have precluded the formation of atomic nuclei. The stability of nuclei would have been adversely affected had the difference in mass between the proton and electron not been about equal to twice the mass of the electron. Hoyle has emphasized that unless certain atomic nuclear resonances were at precisely the correct frequencies to permit the formation and persistence of carbon nuclei inside stars, carbon would not be produced efficiently, and without an adequate supply of carbon, life as we know it would not be possible. Moreover, the transformation of carbon to oxygen by incorporation of a helium nucleus is moderated by a resonance in the oxygen nucleus, so that the carbon formed does not all disappear. Hoyle (quoted by Davies, 1982) finds the whole situation so amazing that he has said: 'a commonsense interpretation of the facts suggests that a superintellect has monkeyed with physics, as well as chemistry and biology, and that there are no blind forces worth speaking about in nature'. It has been suggested that a vast number of universes might exist, and that ours is a rare example of just the right constitution to generate life.

These profound problems are of fundamental importance but for our present purposes we shall have to concentrate on the possible modes of origin of organisms, given the type of Universe that developed after the Big Bang. As was mentioned earlier, some scientists believe that life must have originated somewhere other than on Earth, but wherever it arose certain developmental sequences can reasonably be considered as probably necessary. Usually, the sequence is envisaged in the form referred to by Bernal (1967) as (a) from atom to molecule; (b) from molecule to large molecules – polymers; (c) from polymer to organism. It is common practice to speak of *chemical evolution* during the pre-biological stage of development, when the existing conditions and the nature of the reactants would have favoured the production of some molecules rather than others. Evolution and selection in the Darwinian or neo-Darwinian sense requires the development of replicating systems, such as those containing nucleic acids, with a capacity to generate copies of themselves, so that rare chance variations in structure (as in sequence of bases) can be multiplied and become common.

Pirie (1954) proposed that the process of production of living from non-living matter should be called *biopœsis*, and we shall use this term. We now know that living organisms were present on Earth as long as 3.5 billion (even perhaps 3.8 billion) years ago, and that the Earth is about 4.5 billion years old. The production of prokaryotic organisms, if it occurred on Earth, seems therefore to have been fairly rapid. Assuming that life arose here, was it the result of some isolated and highly improbable association of molecules at one place, or was it a straightforward consequence of the way atoms and molecules behave under the conditions that existed? Is life on Earth an example of an exceedingly rare phenomenon in the cosmos, or shall we be able to conclude that other planets, more or less similar to the Earth, if they exist, are probably harbouring organisms of some sort?

At the meeting of the British Association in Dundee, in 1912, E. A. Schäfer spoke of the origin of life from the evolutionary standpoint:

> Looking, therefore, at the evolution of living matter by the light which is shed upon it by the study of the evolution of matter in general, we are led to regard it as having been produced, not by a sudden alteration, whether exerted by a natural or supernatural agency, but by a gradual process of change from material which was lifeless, through material on the borderland between the animate and the inanimate to material which has all the characteristics to which we attach the term 'life'.

Schäfer emphasized that, at that time, there was no certain knowledge of the mode of this transformation. At about the same time, a short book on the origin of life was written by Benjamin Moore (1913), Professor of Biochemistry at Liverpool. Moore wrote that:

> It was no fortuitous combination of chances, and no cosmic dust, which brought life to the womb of our ancient mother earth in the far distant Palaeozoic ages, but a well-regulated orderly development, which comes to every mother earth in the Universe in the maturity of her creation when the conditions arrive within suitable limits.

The resurgence during the 1920s, 1930s and in the post Second World War years, of interest in the problem of the origin of life was much influenced by statements of this kind, and many (the optimists in the field) would still regard them as essentially correct. In recent years, a more pessimistic group has emerged, oppressed by what they consider to be the improbability of life development elsewhere. Others, including Hoyle, believe that life is widespread in the Universe, but did not develop spontaneously on Earth.

The theories of Haldane and Oparin

During the 1920s, two scientists, J. B. S. Haldane in England and A. I. Oparin in the USSR, were thinking and writing about the origin of life. Working independently, they arrived at rather similar and in some ways complementary conclusions, so that the term 'Haldane-Oparin Theory' has been applied to their views.

Haldane's (1929) essay on the origin of life contains a summary of his position. He suggested that the Earth's primitive atmosphere probably contained little or no oxygen, because the present supply of oxygen is only about the amount necessary to burn all the coal and other organic remains below the surface. On the other hand, he thought that there was probably much more carbon dioxide in the primitive atmosphere than there is now, and that much of the nitrogen now free in the atmosphere was combined as nitride in the Earth's crust. Free ammonia would have been produced by the action of water on metal nitrides. At the present time, much of the solar short-wave ultra-violet radiation filtered off by ozone (a form of oxygen, O_3) in the atmosphere. Before there was much free oxygen, this ozone layer would have been almost absent, so that more ultra-violet radiation would have reached the Earth's surface.

Even before 1929, when Haldane's essay appeared, Baly, of Liverpool, had shown that when ultra-violet radiations act on a mixture of water, carbon dioxide and ammonia, simple organic chemicals may be formed, including probably amino-acids. Haldane, therefore, had some experimental backing for his further suggestion that similar processes in the primitive atmosphere might have led to the accumulation of organic compounds in the oceans, until they 'reached the consistency of hot, dilute soup'. Whereas, today, an organism must compete with others for food, the precursors of life which might have formed under these conditions would have been in a rich chemical soup and would have found 'food' all around them. Little or no free oxygen would have been available to them.

Many organisms exist which can live in the absence of free oxygen, and indeed some are unable to live in its presence. Organisms needing to live without oxygen are called *anaerobes*. Some organisms can live in the presence or absence of oxygen. As some derivatives and forms of oxygen are toxic to organisms, those utilizing oxygen have to develop protective mechanisms, enzymes that inactivate the poisons. *Aerobic* organisms, able to utilize oxygen to oxidize nutrients, can obtain more energy from a given quantity of food than anaerobes. Nursall (1959a, 1959b) suggested that the accumulation of free oxygen in the atmosphere (after the development of oxygen-releasing photosynthetic mechanisms) was a prerequisite for the origin of many-celled animals (metazoa).

Oparin first expressed his views on the origin of life at a meeting of the Russian Botanical Society in 1922, and a short account of them was published in 1924. Later, he developed his theories at greater length in publications available in English translation (1957, 1961, 1964, 1968). Haldane had thought that carbon dioxide (CO_2) was the most likely form in which carbon was present in the primitive atmosphere, but Oparin suggested that methane (CH_4) was more probable. Current thought would favour Haldane's CO_2. Haldane and Oparin accepted the view that free oxygen has accumulated in our atmosphere largely as a result of photosynthesis.

Oparin attempted to outline a possible mode of transformation of organic chemicals accumulating in the oceans into simple, localized systems from which the earliest cells could have developed. The cytoplasm of present-day cells is bounded by a double layer of lipid molecules (*lipid bilayer*) with associated proteins, the cytoplasmic membrane, and the structure of this membrane, confer on it the powers of selective uptake of nutrients and discharge of wastes, both complex and essential for cell functions and survival. Oparin suggested that the modern type of cell might have been preceded by something simpler. He pointed out that, in solution in water, organic chemicals do not necessarily remain uniformly dispersed but may form layers and droplets. The formation of one kind of aggregate of fairly large molecules in water, so-called *coacervates*, had been studied by Bungenberg de Jong (1931–2). Such droplets, rich in organic constituents, may separate from solutions, and the droplets are probably surrounded by a tight 'skin' of water molecules. They may join with each other but do not disperse into the surrounding water. These droplets can display some features reminiscent of the behaviour of living cells, and do lead to compartmentalization of certain organic molecules. Oparin argued that as organic compounds collected in the oceans, coacervates could have formed, and the constitution of these would have varied. Among the kinds formed, some would have been better able to persist than others, and by a sort of physico-chemical selection would have come to predominate. In the laboratory, it is possible to incorporate enzymes into coacervates. The significance, if any, of coacervates for the origin of organisms is not known, but they do demonstrate one way in which certain types of molecule can be brought into association and localized. Any resemblance they have to living cells is at best superficial.

In a later article, Haldane (1954) gave a useful classification of theories of the origin of life, which may be summarized as follows:

1. Life has no origin. Matter and life have always existed.
2. Life originated on our planet by a supernatural event.
3. Life originated from ordinary chemical reactions by a slow evolutionary process.
4. Life originated as a result of a very 'improbable' event, which however was almost certain to happen given sufficient time, and sufficient matter of suitable composition in a suitable state.

The first suggestion is incompatible with the now widely accepted view that the Universe originated in a Big Bang, and so is of finite age. It would be consistent with a steady-state universe. What we might call the probiological optimism of the first two decades following the Second World War has been tempered by the negative findings of space probes and by an increasing realization of the functional complexity of even simple organisms, so that a greater proportion of scientists than formerly would probably tend to favour 4 over 3 at the present day; but it would not be surprising if new ideas derived from the study of dissipative structures were to lead in the not too distant future to a reassessment of the notion of 'improbability' in this field. If we imagine a number of similar planets like the Earth, life would appear on all of them after roughly the same time if possibility 3 were true. If, however, one stage in the production of living things has a very low probability of occurrence, there might be large differences in the time of appearance of organisms on the individual planets, and some might never bear life.

Biological membranes

Biological membranes, including those forming the boundaries of cells and as important components of intracellular structures, such as mitochondria and chloroplasts, consist of complex arrangements of protein and lipid molecules and are of fundamental significance in present-day organisms. The question of how the components of cells became membrane-bounded is therefore important. Goldacre (1958) studied surface phospholipid/protein films (derived from plant material) which occur on lakes and rivers, and noted that deformation of the surface, as by winds, and reduction in surface area caused by alterations in depth and flow characteristics, led to crinkling and folding of the films and consequent formation of folds containing alternately air and water. Escape of air leads to collapse of air-containing folds and fusion of parts of the film to produce vesicles containing water and bounded by a double lipoprotein membrane. If phospholipids and protein were formed on the primitive Earth, as some investigators believe, a similar mechanism could have played a rôle in providing a means of compartmentaliza-tion of other prebiotic molecular systems. The chemical structure of lipids and proteins favours the production of ordered molecular arrays under suitable circumstances, as at air/water interfaces. (See Deamer and Burchfield, 1982.)

It should always be borne in mind that the formation of structured molecular arrays occurs under various circumstances, and must have been crucial for biopœsis, wherever life developed. The functioning of living cells depends not merely on having the right components present but on their mode of organization. A factory consisting merely of a pile of machine components would not function. At the molecular level, significant types of self-organization occur, and may have helped to bring about the development of compartmentalized systems on which natural selection could operate.

In the absence of replicative processes, true natural selection can not be said to occur, but during the stages of chemical evolution, some structures would prove to be more stable than others and so persist. Oparin suggested that in the course of prebiological and early biological evolution, intermediate forms linking 'animate' and 'inanimate' systems might have been swept away, so that now we feel keenly aware of a 'gulf separating organic and inorganic nature'. We shall see later how the transition from 'inanimate' to 'animate' might have occurred by interaction of inorganic systems and organic molecules, and by the development of replicating molecules or structures.

The origins of organic chemicals

The theories we have so far mentioned in this section postulate the production of organic chemicals under the influence of ultra-violet radiations, and the work of Baly gave some support to this view. In experiments designed to test this hypothesis, it is important to exclude any present-day micro-organisms which might contaminate the mixtures of simple chemicals used as starting materials and, by their activities, produce changes which could be attributed mistakenly to the radiations. More recent work, in both the USA and the USSR, has confirmed the possibility of transformations by ultra-violet radiation, and has shown that other energy sources, too, might have been instrumental in promoting the synthesis of organic compounds on the primitive Earth. Molecules, including those of water, ammonia and simple hydrocarbons, can be split to give highly reactive chemical groups, or *radicals*, which can then react with each other or with other materials to produce new and more complicated compounds.

Electrical discharges in the primitive atmosphere, including lightning and corona discharges from pointed objects, would have been another source of energy. In a series of experiments, Miller (1953, 1955, 1959), working in Urey's laboratory, showed that organic chemicals could be formed in artificial gas mixtures subjected to electrical discharges for about a week. Several different forms of apparatus were constructed for the experiments, and both spark and silent discharges were used. The gas mixtures contained methane, ammonia, water vapour and hydrogen. At the conclusion of the experiments, several amino-acids had been formed, together with numerous other organic chemicals. These substances were not produced if free oxygen was added to the mixture. Different gas mixtures were later investigated by Abelson (1956), who confirmed Miller's work and also showed that amino-acids were still produced if ammonia were replaced by nitrogen and methane by carbon dioxide or carbon monoxide (CO). Further experiments were reported by Pavlovskaya and Pasynskii (1959), who suggested that carbon monoxide might have been an important atmospheric constituent at the time of formation of amino-acids, and that free hydrogen was not essential. Many workers now believe that free hydrogen was lost rapidly from the primitive atmosphere and that, as Calvin had suggested, carbon dioxide was a major constituent.

Radioactive compounds on the primitive Earth would have been a source of alpha-, beta- and gamma-radiation, and there is evidence that these energy sources might have played a rôle in organic chemical formation.

Alpha-particles

An alpha-particle is a helium nucleus and thus consists of 2 neutrons and 2 protons. Several instances have been reported of the formation of organic compounds from carbon dioxide and other substances under bombardment by alpha-particles. In 1926, Lind and Bardwell obtained resinous organic substances by alpha-particle irradiation of mixtures of carbon dioxide, carbon monoxide and hydrogen or methane. Sokolov (1937) suggested that water in sedimentary formations would be split by alpha-rays to give hydrogen and oxygen. The oxygen would be removed by reaction with metals and any existing organic compounds, and the hydrogen might then lead to the production of methane from carbon dioxide. The methane so formed could be the starting point for the formation of hydrocarbons of higher molecular weight. Garrison (*et al.*, 1951) and his colleagues showed that organic compounds, including formic acid and formaldehyde, were formed when solutions of carbon dioxide in water were exposed to alpha-particles in a cyclotron.

Beta-particle bombardment

It was mentioned earlier that beta-rays are streams of electrons. Hasselstrom (*et al.*, 1957) and his colleagues found that bombardment of a solution of ammonium acetate by beta-particles promoted the formation of two amino-acids (glycine and aspartic acids), and a third unidentified amino-acid. It was suggested that the beta-particles were knocking out hydrogen atoms from the compound and that the resulting radicals reacted to form the products observed.

Gamma-radiation

Exposure of ammonium carbonate to gamma-rays from radioactive cobalt has been shown to lead to the formation of organic compounds, including glycine. Paschke (*et al.*, 1957) has suggested that gamma-radiation from terrestrial sources might have been more important than sunlight or electrical discharges for the production of organic chemicals on the primitive Earth.

It is clear that reactions of the type mentioned might have played a rôle in the formation and transformation of organic compounds on the primitive Earth, assuming that conditions were such that the compounds formed could persist and were not broken down by excessive radiation levels or by reaction with other materials, including oxygen. Oparin (1957) suggested that the main role of sources of energy, including cosmic rays, emissions from terrestrial radioactive sources and electrical discharges, was the promotion of reactions among simple hydrocarbons which were, in his opinion, the most plentiful carbon compounds at that time. In support of his view, Oparin quoted spectroscopic evidence from studies of stars, comets and planets which suggested that the production of hydrocarbons is widespread throughout the Universe.

The primitive atmosphere. Further considerations

We have seen that Haldane and Calvin suggested that carbon was present in the primitive atmosphere mainly as carbon dioxide, CO_2, whereas Oparin favoured the presence of carbon in a chemically reduced form, as methane, CH_4. Vinogradov (1959) argued on geological and geophysical grounds that carbon was probably present as carbon monoxide, CO. He was led to this conclusion by a consideration of the composition of volcanic gases, which contain water vapour, carbon dioxide, carbon monoxide and a number of other constituents, but methane is found, if at all, in only small amounts. Goldschmidt (1952) stated that 'without doubt' the main carbon compound of the primitive atmosphere was carbon dioxide, a gas which has been discharged from the interior of the Earth during the whole of geological history. There is fairly general agreement that the primitive atmosphere contained only small amounts of free oxygen.

In a valuable recent review, Gillett (1985) has stated that the view that the primitive atmosphere was a strongly reducing one of methane, CH_4, and ammonia, NH_3 has become increasingly discredited. Outgassing from volcanic sources would not have provided a methane-ammonia atmosphere. In contact with hot rocks, the hydrogen of methane would have reacted with the oxygen of oxides in the rock to give water, H_2O, and the carbon would have given carbon monoxide, CO. Moreover, a methane-ammonia atmosphere, had it existed, would in fact have been decomposed by ultra-violet radiation and the hydrogen freed from combination would have escaped from the Earth's atmosphere. The carbon of methane should then have been deposited as graphite, but deposits of appropriate age are not found.

Present-day views on movements of the Earth's surface materials (plate tectonics) involve the concept of passage of sea water, carrying dissolved gases, into the Earth's mantle, which would also tend to lead to the destruction of a methane-ammonia atmosphere. All of this is in keeping with views that the primitive atmosphere was less reducing than Oparin supposed, and probably contained more carbon dioxide than the present-day atmosphere. In the remote past, the Sun was less luminous than it is now, and the excess carbon dioxide, by its greenhouse effect, would have helped to keep the Earth warm. In the course of time, the carbon dioxide reacted with rocks to form carbonates, and carbonate rocks are present in ancient formations. It is quite possible that production of some free oxygen by photolysis of water in the atmosphere led to the production of ozone at a fairly early stage in the Earth's history.

On this view, the starting materials for life were compounds such as formaldehyde, CH_2O, and cyanide, HCN, but other possibilities still remain. Recent explorations of the sea bed have revealed the presence of unusual local conditions where hot vents (named *black smokers*) occur. At these points, sea water reacts with hot volcanic rock, and under these circumstances methane and ammonia can be produced. If syntheses of the type investigated by Miller were important for the genesis of the building blocks of life, events in the oceans might have been more significant for their production than the state of the atmosphere.

The atmosphere and biopœsis

Many investigators and speculators on the origin of life have concluded that oceans probably played a crucial rôle in the early stages of biopœsis, after some synthesis of organic chemicals had occurred in the atmosphere. The surface waters of the Earth became the primordial soup as compounds were transferred from the atmosphere. In the Haldane-Oparin theory, the precursors of cells were considered to be coacervates, or something similar. The first organisms used the organic materials of the soup, and were therefore *heterotrophs*, requiring a supply of complex nutrients. Photosynthesis was postulated as a later development after depletion of the 'foodstuffs' in the ocean.

The shortcomings of this scheme have been forcibly pointed out by Carl Woese (1980), who stressed the fact that basic biochemical reactions involve the removal of water (dehydration – as, for example, in the formation of peptide bonds), and considered it unlikely that they would have begun in an ocean. He also pleaded for a more holistic view of biology, recognizing that the precursors of life and living organisms should share common organizational principles, as both are concerned with the transformation of energy into organization. It is not then unreasonable to suggest that visible light was the primary source of energy for life, as it is now through the photosynthetic activities of plants, and that photosynthesis of some type was important from the beginning. The early organisms, rather than simply using up materials made by Haldane-Oparin processes, were probably *autotrophs* (able to live on simple chemicals such as CO_2 and ammonium salts), and photosynthesizers, so that they contributed to the production of complex molecules, and were sources of organization.

In support of his view, Woese quotes evidence that even as long as 4 billion years ago there may have been organisms on the Earth, yet the Earth was formed only 4.5 billion years ago. He points out that if the origin of life practically coincided with the formation of our planet, it is difficult to regard it as an improbable event, or an event requiring prolonged accumulation of chemicals in an ocean. It is also likely that the Earth's surface was hot, so that there would have been no ocean, but water would have been dissolved in molten silicates. Indeed, taking into account its present temperature, the distance of the Earth from the Sun seems to be just about great enough to prevent the development of a runaway-greenhouse effect of the type which has always prevented Venus from having an ocean. It seems probable that when the Earth was hotter, a runaway-greenhouse condition did develop. The atmosphere, in which water would condense on dust particles to form droplets with dissolved salts, would have been 'a stratified, steady-state, chemical and photochemical system, under severe, highly reactive conditions'. The droplets could play the rôle of protocells, so that all stages of evolution might have been 'cellular', with important chemical reactions taking place at interfaces, whose total surface area would have been enormous. The concentration of light-absorbing molecules, acting as transducers of electromagnetic to chemical energy, could have stabilized the droplets. This scheme does not require the intervention of ultra-violet light or electromagnetic discharges to drive chemical reactions. It raises the possibility that, as others have from time to time suggested, the dense atmospheres of planets such as Venus and Jupiter might provide useful clues to prebiotic chemistry.

In Woese's scheme, the runaway-greenhouse effect has to be broken to permit the formation of oceans and the type of Earth we now know. Moderation of some sources of heat, including meteorite impact and radioactivity, would have led to some cooling, but biological or prebiotic activity could have been important, too. One group of prokaryotic organisms, the *methanogens* (*Archaebacteria*), is considered to be of great antiquity, and ability to produce methane (CH_4) by reduction of carbon dioxide and carbon monoxide may have been a feature even of entities that could be considered prebiotic. This could have led to a drop in the concentration of CO_2 and CO in the atmosphere to levels which would no longer maintain the runaway-greenhouse condition, so leading to further temperature reduction and permitting the formation of oceans. To make this possible, evolution in the primitive atmosphere would have had to advance to the stage of methanogenic systems, probably including a mechanism for protein synthesis and even a simple genetic system.

As with other schemes for the genesis of prebiotic and early biological systems, the successful production of organisms as we know them today would depend on the concurrence of appropriate values of many different parameters, and the 'improbability' of generating organisms appears to lie in the chance of getting everything right at the same time, rather than in difficulties of visualizing biopœsis under 'ideal' conditions. Even if there are many millions of planets basically similar to the Earth, small variations in other conditions, distance from parent body, temperature, luminosity of parent body, etc., might preclude biopœsis, added to which we have to consider the basic uncertainty about the direction that development might take at critical 'bifurcation' points in systems far from equilibrium.

The gas *ozone* which now filters off much of the solar short-wave ultra-violet radiation is an *allotropic* form of oxygen. Whereas the ordinary oxygen molecule consists of two oxygen atoms combined together – O_2 – the ozone molecule is made up of three oxygen atoms, O_3. Ozone can be formed by exposing oxygen to ultra-violet radiations, or to electrical discharges. In our atmosphere, it occurs mainly between 10 and 30 miles above the Earth. If all the ozone in the atmosphere could be collected and brought back to sea level at 0°C, it would form a layer only about 3 mm thick. A layer of ozone of this thickness is said to be an absorbing medium for ultra-violet radiation equivalent to a 500 m thick layer of carbon dioxide, or 10 m of clear water.

The surface of the primitive Earth would not have been completely unscreened against solar ultra-violet radiation, as products of gaseous dissociation would have exerted some screening effect, and very small quantities of ozone might have formed from oxygen released by splitting of water molecules by ultra-violet radiations. Consequently, there was probably some grading of wavelengths of radiation reaching progressively deeper layers of the atmosphere and this might have helped to determine the types of reaction possible at different levels.

Amino-acid mixtures and protein formation

There seems to be no great difficulty in accounting for the formation of a variety of compounds containing chains of a few carbon atoms on the primitive Earth. In addition, there is the likelihood of acquisition of many types of carbonaceous molecules from inter-stellar dust. It is not unreasonable to suppose that similar stages in chemical evolution and molecular acquisition would have occurred on other similar planets in the Universe. Events of this kind represent, however, only a very small step in the direction of life, and the serious problem of the production of systems of great complexity from a seemingly chaotic background has to be faced.

Amino-acids may be regarded as the 'building blocks' of the much more complicated peptides and proteins. Proteins are found in all living organisms known to us and play such fundamental rôles that it is difficult to imagine that carbon-based organisms not containing proteins could exist. Since a number of experiments with somewhat different starting materials have demonstrated the possibility of the formation of amino-acids, the basic components of protein mole-cules were probably produced at a quite early stage. Fox has noted that in several different experiments, organic compounds in the same rather narrow range of possibilities have been formed. It has also been demonstrated, particularly by Fox (1956, 1959, 1977) and his colleagues, that protein-like molecules may be produced in amino-acid mixtures in short periods of time, provided that certain conditions or proportions of different amino-acids, and temperature, are met.

Thermal synthesis of proteinoids

The investigations of Fox have shed light on the mechanism of formation of protein-like molecules (*proteinoids*) in amino-acid mixtures. The amino-acid components of proteins are linked by peptide bonds, and it can be seen from Fig. 21 that the formation of a peptide bond involves, in effect, removal of water. It seemed possible that peptide bonds might form in heated amino-acid mixtures if the temperature was high enough to drive off the water produced in the reaction. Earlier work on the heating of amino-acid mixtures, however, gave little cause for hope, as only biologically useless, tarry materials have been formed.

69

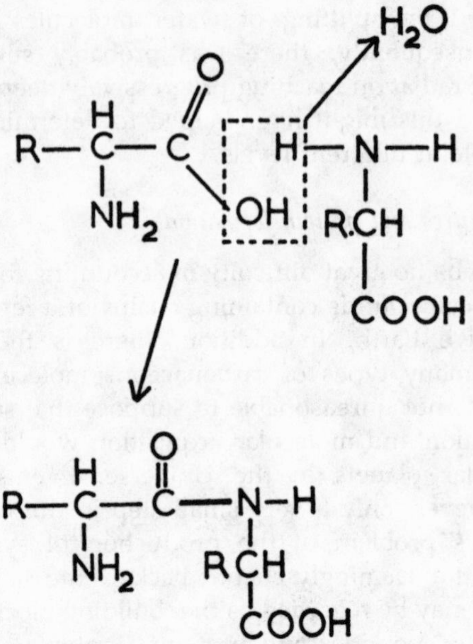

Fig. 21 Water removed in peptide bond formation

Fox and his colleagues were impressed by the fact that, of the twenty or so amino-acids found in proteins, two, known as glutamic acid and aspartic acid, were present in greater amounts than the others. These two constituted about half of a typical protein molecule, and it is known that they play key rôles in biological processes. Harada and Fox therefore tried heating a mixture of amino-acids in which glutamic and aspartic acids were present in excess, and they found that after three hours heating at about 18°C, a proteinoid had been produced. In a number of experiments, proteinoids with molecular weights ranging from 3000 to 10,000 were formed, and they contained some of each of the amino-acids of the starting mixtures.

Fox has pointed out that the reactions in heated amino-acid mixtures need not be solid reactions, because, when heated, glutamic acid becomes liquid at the temperatures used (160–80°C). It was shown, too, that the inclusion of phosphoric acid facilitated the process, and this also provides a liquid phase. The sequence of reactions which occurred bore a close similarity to some found in living organisms, and Fox quoted features which were first found in thermal synthesis experiments and later discovered in biochemical mechanisms.

In further experiments, Fox, Harada and Kendrick (1959) showed that hot saturated solutions of proteinoids gave rise on cooling to huge numbers of uniform, microscopic, relatively firm and elastic spherules. The diameter of the spherules could be influenced by the conditions of preparation (e.g., salt concentration), and varied from about 1 to 80 micrometres. Spherules were produced by heating 15 mg of proteinoid in 3 ml of sea water, boiling for one minute and then cooling to room temperature. The spherules retained their form for several weeks. In more recent studies, it has been demonstrated that spherules examined by electron microscopy show a boundary which is double-layered, superficially resembling the appearance of a cell membrane. Chemically, however, it is unlike a cell membrane, which contains much lipid, and has been likened rather to an impermeable cell wall or spore coat.

Other spherules, produced by different means, have been observed by Bahadur, who optimistically named them *jeewanu*, from the Sanskrit word for 'seeds of life'.

The proteinoid spherules of Fox have, like Oparin's coacervates and Bahadur's jeewanu, been hailed as candidates for the rôle of *protocells*, but many difficulties with this view dictate caution. It is not certain that the concentration of organic molecules in the primordial soup would have been great enough to promote the formation of such structures, and it is by no means clear how these entities could have acquired the characteristics of true cells. The degree of organization of the components of a functioning cell is far removed from the simplicity of the proteinoid spherules, and like the coacervates, they lack the machinery for replication in a sense that would make a biological type of evolution possible. Whether or not spherules or coacervates could have incorporated the necessary machinery of nucleic acid molecules and functional proteins if that had been formed separately (e.g., *hypercycles*: see below) is a question that needs further attention. Spherules do demonstrate that the production of structured molecular arrangements of biologically significant molecules can be promoted by fairly simple means. The coacervates studied experimentally have, on the other hand, all been produced from complex starting materials of biological origin, and are not as stable as proteinoid spherules.

Possible rôle of clay and other minerals

Bernal (1951, 1967) suggested that many crystalline substances, such as quartz and aluminium silicate clays, were probably present in the primitive oceans. It is known that organic molecules may become stuck, or *adsorbed*, to the surfaces of particles of this kind, so that clay and sand might have helped to bring together the simpler molecules which could then react more easily to form complicated compounds. In addition, concentration might have occurred in spaces between the particles, even without adsorption. Mineral particles might act not only as passive traps but also as catalysts. Bernal stressed the importance of a medium free from turbulence, in which the diffusion of small molecules would be restricted so that a considerable concentration could be built up. These conditions might have existed in mud beds, under water, on land or in regions alternately wet and dry, as in tidal estuaries. After larger molecules had been formed, the need for mineral support would have diminished, as the new large molecules would themselves perform the function for which the mineral particles were formerly necessary. The new molecules might be the starting point for the production of coacervates or other bounded protocellular structures.

Mechanisms of this kind might have permitted the development of a primitive 'protoplankton' without the necessity for an 'organic soup' as the sort envisaged by Haldane and Oparin. It is possible that the presence of adsorbents, including clay, would have prevented the formation of soups of chemicals. Indeed, as pointed out by Abelson (1956), the purely random reaction of organic molecules in watery solution tends to produce unusable, tar-like masses. It is therefore probable that mineral catalysis in some form were essential to the early stages of biopœsis. The concentration and directing influence of minerals could have influenced profoundly the course of chemical evolution.

In discussing the importance of inorganic polymers for the production of organisms, Cairns-Smith (see, for example, his 1971 and 1985 publications) has argued that it is necessary to go beyond Bernal's suggestion that clays helped to build up the organic molecules from which organisms could be formed. The key components of the first organisms might have been inorganic, and not organic polymers. What we now recognize as organisms might have been preceded by inorganic systems with clays as pre-proteins and even inorganic genes, and such a system would have been subject to natural selection. The mineral system might have evolved the capacity to control organic reactions, and in the course of time the 'organisms of the first kind', the inorganic systems, would have been replaced by protein-nucleic acid organisms. Cairns-Smith has termed this postulated take-over the 'nucleoprotein revolution'. In support of his suggestion, he points out that nucleotides, of which nucleic acids are built, have not so far been produced in the laboratory under plausible simulated prevital conditions, and that in a primitive soup nucleotides would probably not join to form nucleic acids without the assistance of proteins acting enzymatically (but see reference to Orgel's work on zinc catalysis, below). Moreover, organic minerals are usually tarry and not conducive to biological development. The physical and chemical characteristics of clays promote formation of complexes with other molecules and catalysis of organic reactions.

Andrew (1979) has argued that the 'first organism should have been of such a nature that current life could plausibly have developed from it by progressive chemical evolution'. The precursor of life might, he suggests, have been a fine floc of precipitated calcium phosphate with an adsorbed thin hydrocarbon film, near the ocean surface, able to trap metal sulphides. This system would consist of the phosphate and sulphides, the hydrocarbons and entrapped water, and would resemble a multi-reactor chemical plant, exchanging materials with the surrounding ocean but retaining higher molecular weight insoluble material. Molecular order, resembling that seen in modern membranes, would be imposed by hydrophilic polypeptides and hydrophobic hydro-carbons. The formation of certain metal-organic molecule complexes could have initiated photosynthetic processes, and have made solar energy available. Structural change, involving rolling up of the originally disk-like system, would give protocells.

Natural selection at the molecular level

In considering the evolution of organisms, Darwin wrote in his book *The Origin of Species*: 'The preservation of favourable individual differences and variations, and the destruction of those which are injurious, I have called Natural Selection or the Survival of the Fittest.' Manfred Eigen (1971), in an important and detailed theoretical treatment of the problem of the self-organization of matter and the evolution of biological macromolecules, pointed out that in order to close the gap between physics and biology 'we have to find out what "selection" means in precise molecular terms which can ultimately be described by quantum-mechanical theory. We have to derive Darwin's principle from the known properties of matter.' As part of this undertaking, Eigen (see Eigen and Winkler, 1982) devised a game based on known properties of RNA. It will be recalled that RNA consists of linked nucleotides, and these are indicated by the labels C, G, A and U, the initials of the base components cytosine, guanine, adenine and uracil. Pairing of nucleotides can occur by the formation of hydrogen bonds between the bases C and G, and A and U. The formation of these cross-links will cause the strand to bend back on itself, giving rise to U-shapes or loops. In the game, each player is given a random sequence of coloured beads, representing the four different nucleotides, and a tetrahedral die, each face of which is of the appropriate colour to represent one of the nucleotides. The players throw the die and, by substituting a position in the sequence with the symbol rolled, attempt to approach a structure with as many as possible GC and AU pairs. The game is played according to certain rules based on known features of the molecules, and ends after a predetermined number of throws or when one player has a completely paired structure. At the beginning of the game, the random base sequence provided in the form of linked coloured beads is examined for chance pairing. The outcome is that for the most rapid construction of a paired RNA structure, three-leaf or four-leaf clover patterns of bending the chain are best for chains of 80 beads (=nucleotides). This molecular shape is in fact found in tRNA, which contains about 80 nucleotides. In this game, the throws of the die can be regarded as 'mutations', and leads to the production of structures with maximal base pairing, representing more stable molecules than were constituted by the original sequence and having the 'advantage' of resistance to breakdown.

An example of what can be regarded as molecular evolution in an experimental situation was demonstrated many years ago by Spiegelman and his colleagues (1970). RNA derived from a bacterial virus could be made to replicate *in vitro* in the presence of the necessary building blocks and a protein enzyme to speed the process. The addition of a 'poison', ethidium bromide, to the system blocked replication, but even in the course of minutes mutant molecules resistant to the poison were produced, and were able to replicate. The significance of such a process for molecular development in prebiological 'soups' (if they existed) is easy to appreciate.

The structure of nucleic acids, leading to the ability to replicate by the formation of complementary strands able to act as templates for production of copies of the original molecule, means that 'successful' molecules in terms of stability and persistence can be copied in large numbers. Proteins, however, are not able to replicate in this way, and production of large numbers of molecules of the same protein now depends on the interpretation, by RNA and the rest of the protein synthesis machinery, of the information stored in coded form in DNA. Many different protein enzymes, whose structures are themselves coded in DNA, contribute to the functioning of the protein synthesis system, and to the production of new nucleic acid molecules, so that we can visualize cyclical processes taking place. DNA and RNA are necessary for protein synthesis, and proteins are necessary for efficient DNA and RNA synthesis. Any theory of the origin of life must take account of the seemingly remarkable interrelatedness of the function of the myriads of molecules organized to carry out these and other functions.

As possible early examples of life-like systems, Eigen (1971) has proposed what he has termed *hypercycles*, based on the assumption that nucleic acids and proteins would have been present on the primitive Earth. Given sufficiently large numbers of molecules in a primordial soup, chance events could lead to minor cyclic inter-actions, with nucleic acid strands replicating by complementary strands mutually reproducing. In the course of time, systems could evolve in which nucleic acids promote the formation of small peptides which themselves are able to catalyse the replication of other nucleic acids, which in turn could code for peptides able to catalyse the replication of yet other nucleic acids. The crucial step would be the production at some stage in the chain of a peptide that would act back on the first nucleic acids in the chain and so close the loop and form a cycle of reactions. Since at each step in the chain, replication cycles are occurring, Eigen refers to the total system as a *hypercycle*. Summarizing the properties of hypercycles, Eigen con-cludes that each system would have *autocatalytic growth* properties, that independent cycles would *compete for selection*, and that the non-linearity of the system would make selection sharp, and the coding and *chirality* ('handedness', relating to the type of optical isomers present) would be conserved. The information capacity would be adapted to the requirements of the system, and the system could evolve by utilization of selective advantages. Ultimately, the system must compartmentalize to permit full utilization of advantageous changes produced by mutation, and may then link its code units into a stable chain, a process that would be facilitated by the evolution of enzymes able to promote joining of the units. Compartmentalization is not an inherent property of hypercycles but, where it occurs, it confers a selective advantage. The hypercycle would have many properties often considered to be characteristic of organisms, and after compartmentalization would constitute a biologically meaningful *protocell*.

In these early stages, the nucleic acid involved was mainly single-stranded RNA. Double-stranded DNA as a store of genetic information is likely to have been a development following this stage, and Eigen has argued that the selection of such informational DNA would have depended on the functional effectiveness of the proteins for which the DNA coded, and not only on molecular stability.

On Eigen's (1971) view, what may be called the origin of life '*turns out to be an inevitable event – provided that favourable conditions of free energy flow are maintained over a sufficiently long period of time. The primary event is not unique.*' Furthermore, the universality of the present-day genetic code would have been guaranteed by the '*sharp non-linear selection procedure*'. Other codes that might have existed would have been ousted by selection. Elsewhere in the Universe, if life exists, different codes based on similar principles might have evolved.

Luria (1973) has stressed the importance of the 'truly creative advance' which occurred when a nucleic acid molecule 'learned' to direct the assembly of a protein which in turn promoted the copying of the nucleic acid itself. Some thirty years ago, in their *critical complex theory*, Eyring and Johnson (1957) postulated a critical event of this kind, and pointed out that replication of the template and the associated crude enzymes taking place in shorter times than their own 'life spans' would have shut the door on competitors with different optical activity. Biological evolution could be considered to have begun at that time.

The problem of self-organization of matter

It will have become apparent that a theme running through discussions of the origin of life is the possibility of the 'self-organization' of matter. Interestingly, as the growing awareness of the complexity of living organisms and the seemingly incredibly vast improbabilities of life arising by 'chance' have impressed more and more investigators, there have been great advances in the appreciation of the possibilities of patterns of organization of matter emerging from what appear at first sight to be unpromising mixtures of starting materials.

More than thirty years ago Pringle (1953, 1954), anticipating more recent theoretical developments, discussed the problem of the localization of chemicals into units from quite dilute solutions of starting materials. He was not strongly impressed with the then popular arguments in favour of a concentrated primordial soup, but suggested that the photolysis of water in the atmosphere by ultra-violet light would have led to a 'rain' of small amounts of oxygen and peroxides. These would have reacted with hydrocarbons formed from metal carbides on the Earth, setting off chemical chain-reactions. To account for concentrations of chemicals formed into restricted localities, which would have been necessary to permit further chemical evolution, Pringle made use of a mathematical demonstration by Turing that certain types of initially homogeneous dynamic systems should undergo a progressive change leading to the appearance of spatial heterogeneity. As has been emphasized, for the occurrence of a true evolutionary process there must be a mechanism which allows competition between units, so that the more 'efficient' survive. Pringle suggested that even in seemingly simple chemical reactions there may be competing routes by which a reaction can proceed, and that in complex reaction mixtures there would be greater possibilities for diversity. This opens up the possibility of a 'biological' approach to the problems of some of these chemical reactions, based on an analogy with population dynamics.

Crick (1982) has expressed the view that there are reasons to believe that the original replicator might have been RNA, and referred to work by Orgel on the replication of RNA *in vitro* with zinc as a catalyst. In the presence of a chain in which all the nucleotides were cytidylic acid (polycytidylic acid), containing cytosine, there was slow production of a complementary chain containing guanine. It was mentioned earlier that many enzymes are involved in DNA and RNA replication, and it is of interest that in present-day enzymes promoting the joining together of nucleotides to form nucleic acids, zinc is present. This might be an example of the amplification of the catalytic function of a metal ion that played an important rôle in prebiological or protobiological processes. Orgel also postulated that in the beginning, catalytic proteins would not have been essential, as RNA could itself have acted catalytically and promoted essential reactions. Some time after this suggestion, it was discovered that one type of RNA found in a unicellular animal did in fact play a catalytic role in its own production, and another example of RNA catalysis has also been reported. Orgel has used the term CITROENS to refer to living organisms, an acronym derived from the first letters of the words Complex Information-Transforming Reproducing Objects that Evolve by Natural Selection. In Orgel's view, all intelligent life on no matter what planet would have evolved from a system of simple CITROENS.

The view that genetic mechanisms in some form must have developed before cells, in order to permit an evolutionary development from simpler to more complex systems, with an increasing capacity to store coded information and act on it, and to replicate with good but not perfect accuracy (to permit mutation and selection) is supported by many investigators. Dillon (1978) has argued strongly in favour of placing more emphasis on the production of the genetic mechanism than on the occurrence of coacervates and similar structures, and for the recognition of viruses as a model for early organisms. In his view, slowly self-replicating chains of amino-acids may have preceded the production of tRNA and mRNA and associated enzymes, and later developments involving the production of DNA and the evolution of the first cells. The properties of living things, Dillon states, result solely from the interaction between the environment and a genetic mechanism, and this mechanism he regards as 'basically the equivalent of life itself'.

Clearly, in the origin of carbon-based life, the coordinated development of nucleic acids and proteins was at some stage important, permitting construction and selection of molecular forms with appropriate activities, and this is only part of the story, as lipids, needed for present-day membranes, as well as carbohydrates, had also to be incorporated into developing systems. Indeed, lipids may have played a role in assisting compartmentalization of protobiological systems quite early in the process of biopœsis, whether it occurred on Earth or on some other planet with more favourable conditions or in comets, as some believe. Conceptually, it is perhaps easier to think of the interactive development of a biosphere, rather than of the development of individual organisms. In this context, the views of Cairns-Smith on the possible importance of clays as systems able to constitute a pre-carbon based life form which in time adsorbed, made use of and 'directed' carbon compounds, and so could help to initiate further development towards carbon-based organisms, deserve close experimental examination. The imperfections which occur in the structure of clays can manifest catalytic activities, and under certain conditions, including association with organic chemicals, clays, which have a layered structure, may separate into sheets which act as patterns for the formation of more of the same clay. 'Mutations' due to imperfect copying could in principle lead to alterations in rates of formation and in catalytic activity, so that an evolutionary process with natural selection could occur. If these views are correct, the mineral 'life forms' he envisages should still exist, as they would not be in direct competition with carbon-based life. Cairns-Smith has suggested that crystals of kaolinite found in some sandstones might be a good starting point for a search for mineral life. Work in Germany has shown that clays having certain catalytic properties apparently can replicate and give rise to subsequent 'generations' of clay with the same activities.

Here it is appropriate to mention the *bioid* concept. It has been argued that a primordial soup could not conserve information but that the most primitive nucleic acid/protein system would already require so much information that its spontaneous appearance in the soup would appear to be incredible. If life arose from soups of some kind, predecessors to the nucleoproteic system seem to be necessary. The term 'bioids' has been proposed for self-organized, primitive metabolic pathways using non-standardized catalysts and accumulating non-coded information, kinetic feedback structures that might have preceded the take-over by nucleic acids and proteins (Decker *et al.*: see Noda, 1978, p. 617). The Hanover Programme in Germany is attempting to find out how the gap between primordial soup and the nucleoproteic system was bridged.

In their discussion of the possibility of the origin of new dynamic states of matter (dissipative structures) Prigogine and Stengers (1984) stress the importance of nonlinearity, instability and fluctuations in these systems that are far from equilibrium. The disruption of a certain state of organization by a fluctuation may bring the

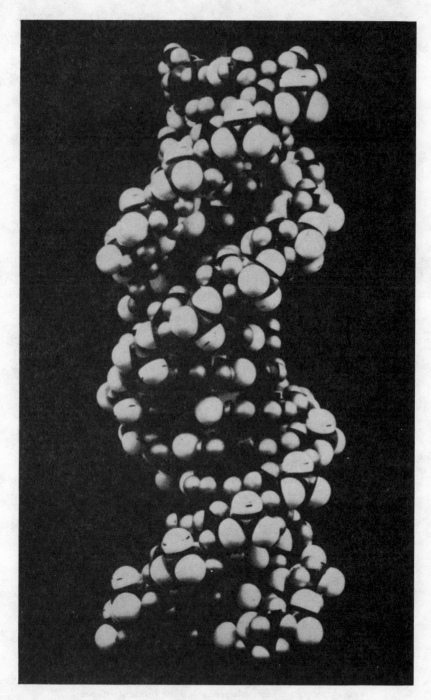

1 Model of DNA molecule, showing helical form. For details of structure see Fig. 13.
Credit: Dr M. H. F. Wilkins, King's College, London

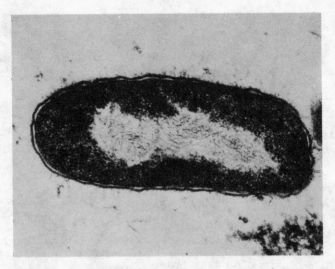

2(a) Electron micrograph of a section through a prokaryote, a bacterium. This type of prokaryote is bounded by two membranes and has a thin wall layer between the membranes. The large pale area in the centre of the cell is the DNA. In prokaryotic cells the DNA is not surrounded by a nuclear membrane. Prokaryotes do not contain mitochondria, but in some ways the whole prokaryote resembles a mitochondrion.

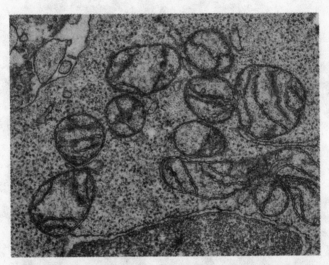

(b) Electron micrograph of a section through mitochondria of a eukaryotic cell. The membrane-bounded structures have membrane extensions projecting into their interiors. Mitochondria contain a loop of DNA, which codes for some of their proteins. Some biologists believe that mitochondria of eukaryotes evolved from prokaryotic cells which associated with and became incorporated into other cells (see p. 151).
Credit (for both): Dr R. L. S. Whitehouse and R. Sherburne, University of Alberta

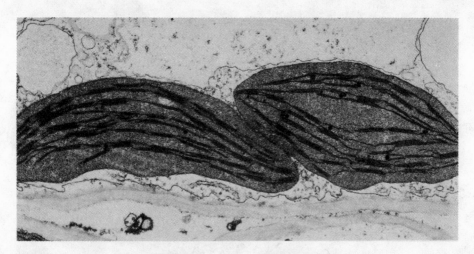

3(a) Electron micrograph of two chloroplasts (sites of photosynthesis) in a higher plant. Complex membrane structures (darkly stained) can be seen inside the chloroplasts. Chloroplasts are considered by some biologists to have evolved from prokaryotic photosynthetic organisms following incorporation of these into other cells to form symbiotic associations. Like prokaryotes and mitochondria, chloroplasts contain nucleic acids.
Credit: Dr J. P. Tewari, University of Alberta

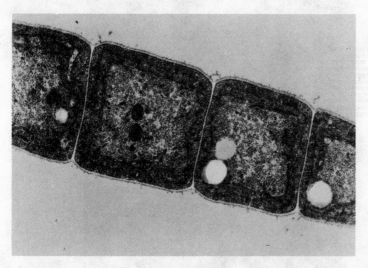

(b) Electron micrograph of a section through cells of a prokaryotic photosynthetic organism. Internal membranes, appearing as dark lines, can be seen inside the cells. Early photosynthetic prokaryotes able to carry out oxygenic photosynthesis are thought by some biologists to have been free-living evolutionary precursors of chloroplasts.
Credit: Professor P. Fay, University College, London

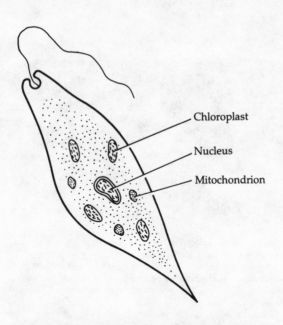

Chloroplast

Nucleus

Mitochondrion

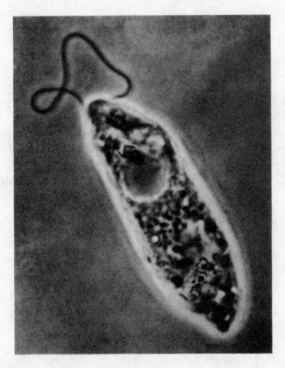

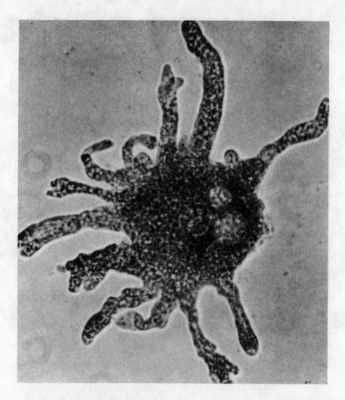

5 *Amœba*. This unicellular organism has the characteristics of an animal cell. It is eukaryotic and contains mitochondria but not chloroplasts, and is unable to use light energy directly. It requires complex foods. The photograph is of a living amœba.
Credit: M. R. Young, National Institute for Medical Research, London

4 (*opposite*) *Euglena gracilis*. This organism, about 1/20 mm long, is unicellular. It is a eukaryote, and contains mitochondria and chloroplasts. In the light, it can carry out photosynthesis. In the dark, it can live on more complex organic molecules, and in this respect resembles typical animal cells. It is sometimes described as a 'plant-animal'. There are other examples of this combination of plant and animal features among micro-organisms.
Credit: M. R. Young, National Institute for Medical Research, London

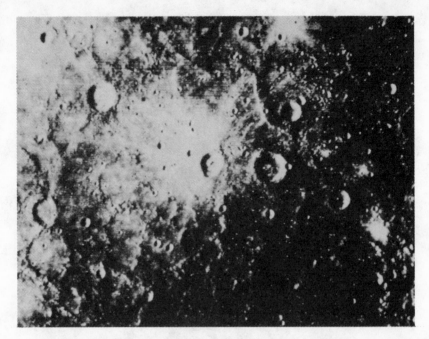

6 The cratered surface of Mercury, a hot, arid and almost certainly lifeless world.
Credit: NASA

7(a) The surface of Mars and part of Viking Lander 1.
Credit: NASA

7(b) The surface of the Moon: the Apennines.
Credit: NASA

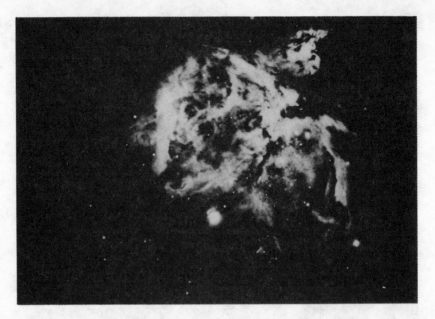

8(a) The Orion Nebula. This appearance is produced by a cloud of interstellar gas associated with bright stars in our own Galaxy. It is known to contain some carbon compounds (including ethyl alcohol!).
Credit: Mount Wilson Observatory

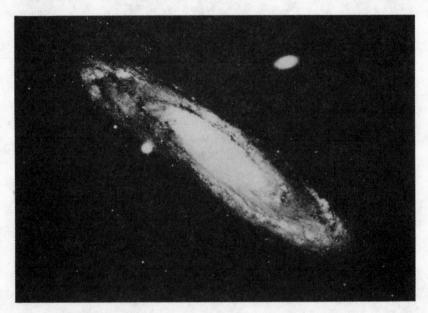

(b) The Andromeda Galaxy. This famous object is a vast system of stars analogous to and separate from our own Galaxy. A fine example of a 'spiral galaxy'.
Credit: Palomar Observatory

system to what they term a 'bifurcation point', at which it is inherently impossible to determine in which direction development will proceed. Chance events may have critical effects and set the system on a new course of development, which is deterministic until a new bifurcation point occurs. In processes of this kind, counter-intuitive outcomes may be observed. The underlying principles may be applicable to a wide range of situations, from inorganic chemistry to sociology. We have mentioned earlier, when considering living organisms, that an overall entropy increase is associated with the localized increase in biological order (which involves local entropy decrease). The irreversible process of energy dissipation is in a sense making possible the localized production and maintenance of organisms.

Taking a broader view, it is possible to argue that the constructive, order-producing effects of entropy increase are in turn dependent on the expansion of the Universe. Hubert Reeves (1985) has emphasized the importance of what he terms the 'Great Extragalactic Elsewhere', in which the entropy can be deposited. As he points out, certain fundamental processes, including the formation of atomic nuclei by aggregation of nucleons, are accompanied by release of energy (the fusion bomb is an example!). In the ordering of atoms as in the formation of ice and crystals, energy is released, and passes into the space created by the expansion of the Universe. As Reeves says, '. . . without expansion, no stable bond could form. . . . Without the elsewhere, no organization could be acquired by matter. The elsewhere is an indispensable precondition for the formation of islands of organized matter, both by reason of the conservation of energy and by reason of its inexorable degradation.'

Theoretical approaches to the problem of 'self-reproduction' have also given fascinating results. The mathematician John von Neumann investigated the possibility of constructing self-reproducing machines, and like Pringle (see above) he was influenced by the work of Alan Turing, who had shown that, if properly programmed, a computer acting according to the few basic rules of logic could perform any conceivable calculation, and if the brain is a complex computer, the appropriately programmed machine should be able to do anything a brain can do. Von Neumann's work has been interpreted as showing that self-reproduction, in the form in which we see it in biology, is not a supraphysical process but could be a feature of a machine. Before many of the details of biological reproduction were discovered, von Neumann's description of a self-reproducing machine contained many features resembling what is now known to go on in living cells. It seems probable that the process of reproduction can be accounted for in mechanistic terms. Moreover, the type of machine

envisaged by von Neumann could make a machine more complicated than itself, and could take part in a process resembling
biological evolution (see Poundstone, 1985). More recently, much
interest has been aroused by the demonstration that 'games'
(perhaps more appropriately 'experiments') can be devised in which
unpredictable patterns are produced (e.g., on a computer screen),
the behaviour of some of the patterns having features resembling
some of those of living organisms. An example is the 'Life' game
invented by J. H. Conway, in which a few simple rules are applied
recursively to generate the observed effects. Speculation on the
significance of these phenomena has ranged widely, even to the
suggestion that entities in some sense 'living' might ultimately be
generated if the process could be extended to a very large scale. To
what extent this might have relevance to the generation of
biospheres remains to be established, but such pursuits are
fascinating and not trivial.

The possibility of building self-reproducing machines, while
relevant in a sense to the problem of the evolution of life, does not
shed direct light on the problem of origins. Any self-reproducing
machine we made would be an extension of human activity.
Computers as we know them do not spring up from mixtures of
components. The same argument would apply to the synthesis of
biological entities by humans in the laboratory. Human intelligence
would have directed the process. We can, however, think of the
various levels of organization in the Universe and imagine a
sequence starting with the Big Bang. Within seconds there would
have been quarks assembled into nucleons and the formation of the
nuclei of hydrogen and helium. In due course, as expansion
continued, stars would form and the heavier elements would be
produced. Along with the birth of stars, clouds of interstellar
organic molecules were probably produced, as carbon and other
elements became available. Planets would form around some stars,
and an occasional planet, with Earthlike qualities, would have an
atmosphere, oceans and become the site of accumulation and
processing of molecules, with new syntheses driven by solar and
other energies. At the molecular level, the theoretical possibilities for
the generation of different combinations are enormous. Initially, an
inorganic (non-carbon based) system (perhaps of clay mineral
'organisms') underwent its own pathway of evolution and interacted
to an increasing extent with organic molecules. This could have led
to the directed development of structures of carbon compounds. At
critical points (the bifurcation points of Prigogine), the course that
development would take would have been unpredictable, and there
would be no guarantee that if we were able to start the process over
again the outcome would be the same. The development of 'bioids'

alongside the mineral systems could also have been important. In time, the compartmentalization of hypercycle-like systems of nucleic acids and proteins would have started biology as we usually think of it, with the evolution of prokaryotic cells, some of which would have developed photosynthetic mechanisms, ultimately achieving the photolysis of water and the release of free oxygen.

Only after a long interval did eukaryotic cells appear, and in time give rise to many-celled organisms, including humans. With the coming of humans and the development of the human brain and its linguistic capabilities, new possibilities arose. Information on the world in which we live was collected, analysed and stored, as in libraries, records, etc., constituting what Karl Popper (see Popper, 1972) has called 'World 3', to distinguish it from 'World 1', the world of material things and 'World 2', the subjective world of minds. At this stage we have made computers and could in principle construct self-reproducing machines, programmed to carry out various tasks, even capable of learning by experience. These could be dispatched into the cosmos on various missions, and would be 'observers' in the general sense of the term. In what sense might such entities be considered to be 'organisms'? Could they be developed to a degree of complexity that would generate consciousness? Is consciousness in fact a manifestation of certain kinds of complexity or something more? Would they ultimately supersede carbon-based life? Is the intuitive feeling many people have correct, that no matter how 'clever' our technological 'offspring' might be, there would be an unbridgable gulf between us and them, that we are in some fundamental way different, or would material differences be compatible with fundamental functional similarities?

Conclusion

On reflection, we may perhaps conclude that, given the type of Universe in which we find ourselves, it is not too surprising that we see order generated and even such seemingly highly improbable things as organisms developing here and there where conditions are favourable. The supreme mystery may be how our critically balanced Universe, permitting us to exist, itself came into existence. Is it but one of a vast number of universes, most of which are sterile? What wonders might other, different universes generate? Is there just one Universe: the one having all physical constants so balanced and quantitated that life can emerge, so allowing the Universe to be observed? Is the process of observation itself creative? Is only one type of Universe possible? Did the Universe arise as a result of a fluctuation in a 'vacuum' and is it, therefore, as one physicist (Alan Guth) has recently suggested, the 'ultimate free lunch', coming literally from *nothing*? Are there as yet unknown directive 'fields' or influences in the Universe, or acting on it, that play a role in the patterning of complexities, as Hoyle implies in his concept of the 'Intelligent Universe'? To all of these questions we can give no certain answer at present.

To borrow a phrase from Sir Bernard Lovell, we find ourselves in the 'centre of immensities'. Physics and biology have undergone great changes since the nineteenth century, and it is hardly surprising that during the few decades of rapid advance it has not been possible to solve all problems. A world built on the principles of classical physics would not function as does our Universe, but the impressive advances in knowledge have been tempered by a heightened realization of the immensities that remain to be understood, and humility and open-mindedness would seem to be appropriate components of the make-up of any scientist (or classicist!). We seem to be living in a Universe that is by its nature *inventive*, as Denbigh has argued. Furthermore, the concepts of 'chance', 'randomness', 'complexity' and 'order' are far from satisfactory, and will probably be refined in the light of further experimental and theoretical discoveries. It is difficult to avoid the conclusion that our present world picture is incomplete in some fundamental respect, and we suspect that future developments in physics will probably far outstrip what has up to now been achieved. That new information on the nature of the Universe and the way in which it came into existence will have important implications for biology, and perhaps especially for the pheno- menon of consciousness, seems highly probable. It is our belief that the truth will continue to be approached most successfully by rigorous scientific investigation and theoretical work, and not by facile lapses into mysticism or 'new' religions. Comforting though such non-scientific pursuits may be, they have a poor record of producing advances in understanding of the Universe. For the present, at least, we have to live with uncertainty.

III MYTHS OR MEN?

Before discussing our neighbour worlds in more detail, it is worth saying a little about past ideas. Speculations have certainly not been lacking, some of which sound decidedly peculiar today!

Ancient ideas

From an early stage in Greek philosophy, it was known that the Earth was a globe; so were the Moon and planets. Therefore, why should not at least some of them be inhabited? The ancients had no means of analysing the atmospheres of the planets, or even of knowing for certain that atmospheres existed at all; neither was there any information about temperatures, water content or surface conditions generally.

The most obvious dwelling-place for 'other men' seemed to be the Moon, which was regarded as cool and earthy. Anaxagoras (500–428 BC) wrote that 'it has an admixture of cold and of earth. It has a surface in some places lofty, in others hollow.' This is by no means a bad description, but the earliest long essay devoted to possible lunar life did not come until about AD 70. This was Plutarch's *De Facie in Orbe Lunæ*, which we have already mentioned.

Plutarch's book contains some science and a great deal of imagination. Though he agreed with Anaxagoras that the Moon is a kind of second Earth, with mountains, valleys and ravines scattered over its surface, he rejected the idea of human inhabitants, and replaced them with 'demons' (not necessarily malevolent). The second part of the essay is in dialogue form, and describes the wanderings of departed souls as they glide between Earth and Moon.

It is not clear how seriously Plutarch himself took these speculations, but the next book to contain an account of lunar life, written by Lucian of Samosata half a century later, was pure satire. It was called the *True History* because, according to the author himself, it was made up of nothing but lies from beginning to end. It describes how a party of sailors passing through the Pillars of Hercules (the Straits of Gibraltar) was caught in a waterspout, and whirled up to the Moon. The travellers arrive in time to take part in a full-scale interplanetary war, thereby anticipating Lord Darth Vader by many centuries. The King of the Moon has quarrelled with the King of the Sun as to who should have prior claim on the planet Venus, and the two armies fight with such allies as cabbage-fowls, horse-griffins, sky-crows and centaurs. . . .

Tales of the seventeenth century

Here we can only mention briefly a few stories, and one of considerable interest was written by Johannes Kepler and published posthumously in 1634. Kepler was the great mathematician who was the first to prove conclusively that the planets move round the Sun in elliptical orbits; he was also a mystic, and his classic novel, *Somnium*, or Dream, was really a barely concealed defence of the Copernican system in the form of a medley of genuine seventeenth-century science and supernatural themes. The hero of the story – Duracotus, a young Icelander – travels to the Moon by Demon power, arranged by his mother Fiolxhilda, who is an accomplished witch. On arrival, he finds a strange world indeed:

> The Moon-dwellers are furry beings, indeed fur makes up the larger part of their bodies, which are spongy, puffy and porous. If a Moon-being is caught unawares by the great heat of the day, his fur is singed and becomes hard and brittle so that in the evening it drops off. . . . Most Moon-dwellers have serpentine bodies, and enjoy living in the moderate heat of the morning or evening sun. . . . All over the Moon lie great masses of acorns. The scales of these cones are scorched by the Sun during the day. Then, in the evening, the cones open and new Moon-dwellers are born.

Equally entertaining is *The Man in the Moone*, written by the Bishop of Hereford, Francis Godwin, and published in 1638. The hero, Domingo Gonzales, travels to the Moon on a raft towed by gansas (wild geese) and encounters all manner of wonders. However, Godwin evidently did not mean his book to be anything more than a good story. The same is true of many other books on the subject produced during the seventeenth and eighteenth centuries.

However, ideas were changing. The true status of the Earth had been realized; instead of being of fundamental importance astronomically, it was a very junior member of the Solar System. Many astronomers of wide repute came to believe that some of the planets, if not all, must be inhabited. The most celebrated of these astronomers was William Herschel.

The nineteenth century

As we have seen, Herschel was not only a magnificent observer, but also a pioneer in the field of stellar astronomy. Yet although in many ways he was far ahead of his time, his ideas about planetary inhabitants sound peculiar today. He believed the habitability of the Moon to be 'an absolute certainty', and he was equally convinced of intelligent life not only on the planets but also inside the Sun, which he believed to have a cool, pleasant region below its hot clouds.

Herschel made no secret of his opinions. His contemporary, Johann Schröter, first of the great lunar observers, believed that certain changes on the Moon indicated vital activity there, while in the mid-nineteenth century the German astronomer, Franz von Paula Gruithuisen, announced the discovery of a true lunar city, with 'dark gigantic ramparts'. Gruithuisen also commented upon the Ashen Light of Venus – the faint visibility of the unilluminated hemisphere. This he believed to be due to vast forest fires lit by the inhabitants to celebrate the selection of a new Government, or perhaps the accession of a new Emperor.

In the popular view, too, the existence of men on the Moon and planets seemed quite likely. In 1835, for instance, many readers of the New York *Sun* were deceived by an elaborate hoax, in which the paper reported that Sir John Herschel (William Herschel's son), working at the Cape of Good Hope with a powerful telescope, had discovered life forms on the Moon ranging from horned sheep to grey pelicans, bat-men, and even 'a strange amphibious creature of a spherical form, which rolled with great velocity across the pebbly beach'. The hoax was exposed before long, but it had a wide circulation. One women's group even wrote to the paper inquiring about the best ways of converting the bat-men to Christianity.

The success of the hoax, combined with the views of such men as Schröter, Gruithuisen and William Herschel, was not as surprising as it might sound. At that time the hostile nature of the Moon and most of the planets was by no means generally appreciated, even by astronomers. It seemed reasonable enough to assume that advanced life forms would evolve on any suitable world, and it was only when scientific progress revealed the unsuitability of the bodies of the Solar System that serious doubts began to creep in. In particular, it became clear that the Moon's atmosphere, far from being dense and breathable, was exceedingly tenuous, and useless for the support of animals. Yet even when the Moon had been crossed off the list of potentially habitable worlds, the planets remained – particularly Mars and Venus.

The twentieth century

Mars, believed by many to be much the most Earth-like of the planets, was of special interest, and as recently as 1900 a prize was offered in France to be awarded to the first person to make contact with beings from another world – the Martians being specifically excluded as being too easy! Then there were the views of Percival Lowell, founder of the observatory at Flagstaff in Arizona which bears his name. Lowell was convinced that the apparent network of 'canals' on Mars, which he believed he had mapped in detail, was due to a planet-wide irrigation system. His book advocating the theory of artificial canals was published in 1906, and was taken very seriously, though many of his contemporaries were openly sceptical. In the same period, the Swedish physicist, chemist and Nobel Prize winner, Svante Arrhenius, was maintaining that Venus was likely to be a world in a state similar to that of the Earth during the Carboniferous period more than 250 million years ago, so that there was probably luxuriant vegetation together with insects and amphibians.

Gradually such ideas fell out of favour, and the advent of space research methods, from 1957, altered the whole picture. The Martian 'canals' were finally banished to the realm of myth, together with Arrhenius's *Cytherean* (see footnote, p. 000) vegetation. We have a great advantage over all our predecessors; where Herschel, Schröter, Lowell and others could only guess, we have the necessary data to permit a more balanced judgement. None the less, it is interesting to observe that among scientists of repute there is still a wide range of speculation about extraterrestrial life, as will become apparent in later sections of this book. Many unanswered questions still remain. The question of life in the Solar System, if not already more or less clarified, has at least become amenable to direct investigation. On basic questions, including the nature of living organisms, the relevance of the development of intelligent machines, the existence of other planetary systems and the possibility of various forms of alien life having a different chemical basis, there is still plenty of room for speculation. We should also note that speculation, even wild speculation, is not incompatible with great scientific ability, and may at times be an indication of a fertile imagination, a faculty not necessarily to be despised.

IV THE MOON

Of all celestial bodies, the Moon can be studied in greatest detail by Earth-based observers. There are two reasons for this. First, it is extremely close: the distance from the Earth (centre to centre) ranges between 356,410 and 406,697 kilometres. Second, it has virtually no atmosphere, so there is nothing to obscure its surface features.

General data

The Moon has a diameter of 3476 kilometres, and a mass 0.0123 that of the Earth. It is therefore smaller and less massive than any of the planets (unless we include Pluto), but it is exceptionally large relative to its primary, and there are grounds for suggesting that it should be regarded as a companion planet rather than a satellite. Its density is 3.342 that of water, appreciably less than that of the Earth, Venus or Mercury and slightly less than that of Mars. Old ideas that it may have broken away from the Earth in the early period of the Solar System have been rejected by most astronomers, and it seems that the two bodies have always been separate. Analyses of the rocks brought back by the American astronauts and the Russian unmanned space-probes have shown that the Earth and the Moon are of about the same age (between 4.5 and 5 thousand million years).

Surface features

Even a small telescope will show a vast amount of detail on the Moon. There are large dark plains miscalled seas or *maria*; there are mountains and peaks, swellings or domes, and crack-like features known as clefts or rills (sometimes spelled *rilles*). The whole lunar scene is dominated by the walled circular structures which are always known as *craters*. These range in size from huge enclosures 300 kilometres or more in diameter down to tiny pits too small to be seen at all from Earth. In form they are basically circular, though often broken and ruined by later structures; they have sunken floors, with walls which rise to only moderate heights above the outer surface, and many of them have central peaks or groups of peaks. The most favoured theory of their origin is that they were produced by meteoric impact, though some observers, including one of the present writers (PM), prefer the theory that the large walled plains are more likely to have been due to internal or endogenic forces.

The 'seas' or maria are extensive; the largest of the regular seas, the *Mare Imbrium* (Sea of Showers) has a diameter of 1300 kilometres, and is 7 kilometres deeper in its centre than at its periphery, reckoning from the mean reference sphere for the Moon. Most of the famous maria form a connected system, though one, the *Mare Crisium* (590 × 460 kilometres), is separate. There is also the separate *Mare Orientale* (Eastern Sea, diameter 965 kilometres) which is a multi-ringed structure, first discovered and named by one of the present writers (PM).* This is the only major mare which extends to the far side of the Moon.

The Moon has *captured* or *synchronous* rotation – which is not surprising; tidal forces over the ages have been responsible, and the same is true for all other large planetary satellites in the Solar System. Because the lunar orbit is not circular, the orbital velocity varies, while the rate of axial rotation does not; therefore the effects, termed *librations* – of which the libration in longitude is the most important – mean that at various times we can examine a total of 59 per cent of the entire surface from Earth-based locations. Before the Space Age, the remaining 41 per cent was averted from the Earth and completely unknown.

The first unmanned probe to send back pictures of the averted regions was Lunik 3, in 1959. Since then, detailed maps of the entire surface have been compiled from the various spacecraft, including the Apollo manned missions, which extended between 1969 and 1972.

There are distinct differences between the visible and averted areas. The far side contains no major maria, but there are large walled plains with relatively light floors, together with one exceptional feature, *Tsiolkovskii*, which has a dark floor and a central

*The name *Mare Orientale* was selected because the visible part of the structure lay on the Moon's eastern limb, as seen from the Earth. Since then, an edict by the International Astronomical Union has reversed lunar east and west, so that the *Eastern* Sea is now on the *western* limb.

peak, and may be intermediate in type between a mare and a conventional crater.

Structure of the Moon

Thanks mainly to the Apollo missions, we now have a reliable idea of the Moon's constitution. There is a loose upper layer, or *regolith*, from 1 to 20 metres deep; this is a breccia (see below) made up of many different materials. Below the regolith is a kilometre-thick layer of shattered bedrock, and then comes a layer of more solid rock going down about 25 kilometres. Beneath this are layers of denser rock, and finally the metal-rich core, which may be from 1000 to 1500 kilometres in diameter. The core is certainly hot; the old picture of a 'cold' Moon has been abandoned. It seems that the crust on the far side of the Moon is thicker than that on the Earth-turned side.

Moonquakes do occur but are very mild by terrestrial standards. There are also occasional meteoric impacts, recorded from time to time until all the surviving Apollo stations on the Moon were shut down on 30 September 1977. There is virtually no overall magnetic field.

All the rocks brought home for analysis are igneous; in age they range between 4.7 million years and only about 3.1 million years. In the lavas, basalts are dominant. In the highlands, there is a great deal of plagioclase or feldspar, a calcium-aluminium silicate. Breccias (complex rocks made up of shattered, crushed and sometimes melted fragments) have been found at every landing site on the Moon.

Lack of atmosphere

With its low escape velocity (2.38 kilometres/second) the Moon would not be expected to retain an appreciable atmosphere, and this has indeed been found to be the case. In the pre-Apollo period it was still thought possible that there might be an atmosphere which would be detectable but we now know that the density is negligible – no more than 2×10^5 molecules per cubic centimetre, which corresponds to what is normally regarded as a high laboratory vacuum. The so-called atmosphere is a collisionless gas in which hydrogen, helium, neon and argon have been identified, but it is certainly not being economical of the truth to say that to all intents and purposes the Moon is an airless world. Occasional local obscurations, seen by all practised lunar observers and known as TLP (Transient Lunar Phenomena), are presumably due to short-lived emissions from below the surface.

Has life ever existed on the Moon?

Even in comparatively modern times there have been suggestions that some form of lunar life might survive; thus W. H. Pickering, who died in 1938, was firmly of the opinion that certain dark patches in some of the craters, such as Eratosthenes, were due to vegetation or even swarms of insects!

In our opinion, the probability of finding indigenous carbon-based lunar life would appear to be practically zero. There seems to be no water on the Moon, even in combination with any of the rocks so far examined, and traces of carbon detected were derived mostly from meteorites or the Sun. The absence of an atmosphere and lack of any evidence of biological modification of surface materials virtually rules out the existence of life at the Moon's surface. The exceedingly remote possibility of some form of microbial life deep beneath the surface, if any water is available, remains, but the chance of biopœsis having occurred at any time on the Moon seems to be miniscule, and it is most improbable that the Moon was ever in a state to permit any of the proposed pathways to life to be followed.

If seeding of the Moon's surface by micro-organisms from space has taken place, there would have been no realistic chance of growth or even of significant survival (except, perhaps, limited survival of hardy bacteria delivered and remaining in space probes).

The seemingly exceedingly remote possibility of the existence of other types of organized matter, such as 'mineral organisms' or 'physical life' on, or in, the Moon may be a matter for consideration at some future time.

Human survival on the Moon

Men have been to the Moon, and were faced with no insuperable technological problems; there is little doubt that the setting up of lunar bases in the foreseeable future would be quite practicable. Living on the Moon will always have to be in artificially constructed controlled environments. The prospects are encouraging, and a lunar base would be of immense value to mankind.

V MARS

The belief that Mars might harbour advanced plant, animal or even intelligent life has a long history, and persisted up to the time of the controlled Viking landings in 1976. Certainly, in many respects Mars appears at first sight to be less hostile than any other world in the Solar System apart from the Earth. Essential data are as follows:

Distance from the Sun	Max. 249.1 million km.
	Mean 227.94 million km.
	Min. 206.7 million km.
Sidereal period	686.98 Earth days
	668.6 Martian days (sols)
Axial rotation period	24 h. 37 m. 22.6 s.
Axial inclination	23°59'
Orbital eccentricity	0.093
Orbital inclination	1°50'59"
Diameter	6787 km.
Density (water = 1)	3.94
Mass (Earth = 1)	0.107
Volume (Earth = 1)	0.150
Escape velocity	5.03 km/sec.

Mars can come closer to the Earth than any other planet apart from Venus. Its minimum distance from us is 58,400,000 km. Southern summer occurs near perihelion; therefore climates in the southern hemisphere of the planet show a wider range of temperature than in the north. The effect is much more marked than with Earth, partly because there are no seas and partly because the Martian orbit is much more eccentric than ours.

Surface features

Telescopically, Mars shows a predominantly ochre-red disk, with white polar caps which vary with the Martian seasons, and dark features which are essentially permanent and which were recorded telescopically by Christiaan Huygens as long ago as 1659. Originally it was believed that the dark areas were seas. When it became clear that Mars cannot support large areas of water, it was generally assumed that the dark areas were depressions – possibly old ocean beds – filled with organic matter which could be termed 'vegetation'. This theory persisted up to the close-range examination of the planet by space-probes, the first of which, the US Mariner 4, by-passed Mars in July 1965. Views such as that of E. J. Öpik, who maintained that the dark areas must be made up of material capable of growing and pushing the ochre 'dust' aside, were incorrect speculations on the causes of changes in surface features.

We now know that the dark areas are nothing more than inorganic albedo features and have no obvious relationship with Martian topography; thus the most prominent of them, known as *Syrtis Major*, is a plateau sloping away quite noticeably at its edges, while another, *Juventæ Fons*, has been described as a box canyon (see Carr, 1981). The ochre coating is due mainly to iron-rich oxides, and in the bright areas, often termed 'deserts', is virtually continuous, while in the dark areas the underlying bedrock shows through. Shifting of dust by winds could produce changes in distribution of colour, and account for phenomena often claimed in the past to be the outcome of biological activity.

The white caps covering the poles are visible with a small telescope under favourable conditions. The southern cap shows a greater range in size than its northern counterpart, because of the greater climatic variations, and in addition the two appear not to be identical in composition. The residual northern cap is thought to be composed of water ice, since solid carbon dioxide would sublimate during the warmest part of the Martian year. The southern cap is colder, and may be composed mainly of carbon dioxide. According to Kieffer and Palluconi (1979), this is because of the greater frequency of dust storms in the far south. Near perihelion, these storms raise quantities of dust into the atmosphere, and this to some extent shields the southern cap from the Sun's radiation – though of course some water ice must be present in the southern cap also.

Many maps of Mars were produced during the pre-Space Age period, and some of them were of great interest (see Moore, 1984); but as all have now been superseded by the Mariner and Viking results, there is no point in describing them further here.

The canals

For many years all theories of Mars were dominated by the great canal controversy. These strange, linear features were first reported in detail in 1877, from Milan, by G. V. Schiaparelli, who used an 8½ inch refractor. He termed them *canali* (Italian for 'channels'), and regarded it as possible that they were artificial waterways, built by intelligent Martians to convey water from the snowy poles to the arid equatorial zone. He also found that some of the canals showed abrupt doubling, so that parallel lines would appear where previously only one had been seen. After 1886 the canals were reported by other observers, and Percival Lowell, who founded a major observatory at Flagstaff, in Arizona, mainly to study Mars, mapped hundreds between 1895 and his death in 1916. Lowell was convinced that intelligent activity was responsible, and even wrote (Lowell, 1906): 'That Mars is inhabited by beings of some sort or other we may consider as certain as it is uncertain what these beings may be.'

However, ideas of this sort met with considerable scepticism, particularly as other observers, using very large telescopes, could see no trace of the canal network. Finally, the space-probe results proved that the canals do not exist; they were due to nothing more than tricks of vision (see Moore, 1977a). Moore (1977b) attempted to correlate the 'Lowell canals' with actual topographical or albedo features, but with a complete lack of success.

Atmosphere

The existence of a Martian atmosphere has never been seriously in doubt, but before the Space Age all estimates of its density and composition were very unreliable (e.g., Dollfus' 1951 estimate of a pressure of 83 millibars and de Vaucouleurs' 1954 estimate of 87 millibars and a predominance of nitrogen). In fact, they have served to emphasize the limitations of Earth-based observations of planets carried out at that time. The findings of planetary probes will be described below.

Earth-based observers often observed clouds, some of which were high-altitude phenomena and thought to be analogous to our cirrus, while others – the 'yellow clouds' such as those seen during the great obscuration of 1911, described in detail by Antoniadi (1916) and others – were attributed to dust storms. The 1911 feature covered a large part of the southern hemisphere and persisted for weeks. Öpik (1950) went so far as to suggest that these phenomena were due to asteroids striking Mars and stirring up the dust, but it was more generally believed that the dust clouds were whipped up from the surface by winds.

Before 1965, the overall impression of Mars was of a world which, although unsuited to higher life forms such as are found on Earth, might be quite capable of supporting organisms of some kind, particularly micro-organisms. It was hoped that the instrument packages carried to Mars by space-probes would give a definite answer to the question of life on Mars. We shall see that this has not in fact happened, and that Mars has provided us with plenty of surprises.

The Mariner and Viking probes

The first successful Mars fly-by probes, Mariners 4, 6 and 7 (1965–69), showed that instead of being no more than gently undulating, Mars is cratered, and at first sight appears more like the Moon than like the Earth. The idea of the dark areas as depressions filled with vegetation was disproved; neither were all the bright regions elevated. Indeed, *Hellas*, which measures 2200 × 1800 kilometres and can sometimes appear as brilliant as a polar cap, turned out to be a deep basin, with a floor 3–4 kilometres below the general surface level; another bright area, *Argyre*, was also a basin, 800 kilometres across.

However, it was with Mariner 9, launched in May 1971, that the best pre-Viking results were obtained. Mariner 9 operated in orbit round Mars from November 1971 until contact was lost in October 1972. When it reached the neighbourhood of Mars a major dust storm was in progress, but before long the dust cleared and the pictures sent back were spectacular. For the first time we were able to study the great volcanoes and the deep canyons.

It is now known that the two hemispheres differ. Much of the southern half of the planet is heavily cratered; there are fewer plains but more craters in the north, and there are several volcanic regions, of which the two most prominent are in the *Tharsis* area (centred on the equator, longitude around 115°E) and *Elysium* (25°N, 210°W). The Tharsis volcanoes include *Ascræus Mons* (height over 14 kilometres) and the comparable *Pavonis Mons* and *Arsia Mons*; all are topped by calderas, of which the largest, that of Arsia Mons, is 140 kilometres in diameter. Even more majestic is *Olympus Mons*, with a height of 25 kilometres, a base of over 600 kilometres across (as against only 120 kilometres for Mauna Loa, the largest terrestrial shield volcano) and a complex summit caldera.

Mariner 9 also showed great systems of valleys and canyons; the longest valley system, *the Valles Marineris*, not far from Tharsis, can be traced to a length of 4000 kilometres, with a maximum width of 200 kilometres and a greatest depth of 6 kilometres below the rim. There is also the *Noctis Labyrinthus*, with canyons from 10 to 20 kilometres wide; this system covers an area of 120,000 square kilometres.

Channels are abundant and seem almost certainly to have been cut out by running water in the remote past. The so-called runoff channels are found in the old, cratered terrain; they are small near their sources but increase downstream, and have tributaries. The fretted channels are wide, steep-walled and flat-floored; they also have tributaries and may intersect craters. Even more important are the outflow channels, which are large and start from well-defined local sources. There seems to be no escape from the conclusion that catastrophic flooding occurred, and 'islands' are obvious.

The evidence for the past existence of running water on the surface, as indicated by the forms of valleys and related features, has raised the spirits of those who are still hopeful of finding Martian life. Water in liquid form is usually considered to be essential for critical stages of biopœsis. The present findings leave open the possibility that, although the planet is now probably lifeless, some stages of biopœsis and even the production of simple organisms might have occurred in past ages. The possibility of dormant life forms being present, although exceedingly remote, deserves future consideration.

In 1976 two Viking spacecraft landed on Mars: No. 1 in June, in the 'Golden Plain' Chryse (latitude 22°N, longitude 47°W) and No. 2 in August, in the plain of Utopia (48°N, 226°W). Both functioned almost flawlessly, and sent back detailed analyses of the surface materials (regolith) and atmosphere, and the attempts to detect living organisms will be outlined below. Further data were obtained from the orbiting sections of the Vikings.

Analyses revealed that the main constituent of the atmosphere is not nitrogen, but carbon dioxide (95 per cent). Nitrogen accounts for 2.7 per cent, argon 1.6 per cent, oxygen 0.13 per cent, carbon monoxide 0.07 per cent and water for 0.03 per cent, and there are traces of neon, krypton, xenon and ozone (Owen *et al.*, 1977). The small amount of ozone does not form an effective screen, so the surface of the planet is exposed to strong ultra-violet radiation. Atmospheric pressure is low, attaining 8.9 millibars on the floor of Hellas, the deepest basin, but is less than 3 millibars at the summit of Olympus Mons. When Viking 1 landed in Chryse, the pressure there was 7 millibars. A decrease of 0.012 mb per sol was subsequently measured, caused by carbon dioxide condensing out of the atmosphere and being deposited on the south polar cap, a seasonal phenomenon. Wind speeds were moderate, not exceeding 20 kilometres per hour, but greater wind speeds can certainly occur, perhaps up to 400 kilometres per hour. It must be remembered, of course, that in this tenuous atmosphere winds will have little force.

Temperatures at the Viking 1 site ranged from −85°C after dawn to a maximum of −31°C near noon; those at the Viking 2 site were similar, though naturally the temperatures at the equator are considerably higher, and may rise well above 0°C.

Analyses of surface material indicated 16 per cent of iron, 15–30 per cent silicon, 2–7 per cent aluminium and 0.25–1.5 per cent of titanium. The general composition seems to be not unlike that of the lunar maria. In all probability the crust extends to a depth of about 200 kilometres; below this is a mantle, and below this again is the core. As with the Moon, no overall magnetic field has been detected.

Even in the moistest region, above the polar caps, water vapour is present to only about 1 per cent of the amount found in the Earth's atmosphere, and other regions have up to ten times less. There is some evidence that moister regions may exist on the planet's surface, perhaps associated with underground ice or even liquid water which may occasionally reach the surface. The Martian atmosphere has no characteristics pointing to modification by biological processes. As Lovelock (1979, 1986) has pointed out, on a life-bearing planet the atmosphere would probably indicate a 'planetary entropy reduction' by its composition. The Earth's atmosphere, for example, is 'wildly anomalous', with its large amount of free oxygen. Lovelock believes that organisms on a planet would certainly use the atmosphere as a 'conveyor belt' for products and raw materials, and shift it away from chemical equilibrium. Even before the Viking landers, Lovelock was able to predict with considerable confidence that Mars would be found to be lifeless. It should be borne in mind that departures from equilibrium in a planetary atmosphere might sometimes be the result of non-biological energy flows, so that while an atmosphere in chemical equilibrium is unlikely to be present on a planet bearing a significant biosphere, non-equilibrium might not necessarily indicate biology. Searches for it could be part of a valuable screening programme for planets of other stars which may be accessible to observation from orbiting laboratories of the future.

In the older literature, much emphasis was put on the occurrence of seasonal colour changes on Mars, including a supposed change of colour of certain zones to a greenish tint, with a 'wave of darkening' spreading towards the equatorial zones from the polar cap in the summer hemisphere. In so far as changes occur, and are not observational illusions, they are now thought to be probably the result of redistribution of dust by the considerable Martian winds, a possibility suggested long ago by V. V. Sharonov. The supposed greenish tint reported by some observers is now widely believed to have been due to a contrast effect, but bluish-green tints have been seen in some photographs of rocks on the surface. Indeed, photographs of a particular rock taken some time apart have been found to show varying greenish patches, superficially resembling the appearances produced by terrestrial lichens. The problem is that colour changes can result from many causes, inorganic as well as organic, and the information available is too incomplete to permit a definite conclusion to be drawn.

The search for life

The Viking instrument packages were designed to carry out a number of experiments aimed at detection of Martian organisms, if such existed. The experiments were inevitably subject to fairly severe constraints because of the limitations of payload and the problem of controlling procedures from Earth. Moreover, the 'biological' experiments were designed on the basis of a number of assumptions, particularly that at least some of the organisms that might be present would behave like Earth organisms and produce recognizable effects on exposure to solutions of nutrient materials. Detailed accounts of the results obtained in the Viking experiments have been published (see Horowitz, 1977) and here we shall be able to give only a condensed account of the findings and to discuss the problems of interpretation.

The landers bore television cameras, and enthusiastic 'pro-lifers' hoped that something recognizable as biological would be seen, but they were to be disappointed. At both sites the landscapes had a discouragingly barren appearance. No plant-like object or even suspicious-looking blob could be seen, much less any animal-like or moving entity. It was noted that some of the rocks might have a blue-green tint and, as mentioned above, observations of changes in pattern of a greenish patch have become for the moment almost the last refuge of those seeking evidence for Martian life. In fairness it must be emphasized that only a minute region of the surface has been examined directly, and there may be small localized niches with more favourable conditions for biological activity.

The Viking organic analysis and 'life-detection' experiments

If any organisms were present on Mars, it was likely that there would be micro-organisms. The remote possibility of the presence of large organisms might have been resolved by the television cameras. The main biological approach was predicated on the assumption that Martian micro-organisms might give results similar to those expected with typical Earth-type micro-organisms. Implicit in this assumption was the expectation that Martian organisms would be carbon-based, and that carbon compounds would be present in the Martian surface materials (somewhat optimistically referred to as 'soil', a word which strictly carries the implication of biological activity in its production). Carbon compounds could be produced by organisms (e.g., by photosynthesis) or, as was expected, could have originated by non-biological syntheses as visualized for the primitive Earth, as well as being collected from space. As part of the general chemical-biological programme, an organic analysis of the surface material or *regolith* would be of value.

Organic analysis

This was achieved by a combination of gas chromatography (GC) and mass spectrometry (MS), methods routinely used in terrestrial scientific investigations. The various compounds present in a mixture can be separated by GC and further identified by MS. It was known that on Earth, results with soils were different from those obtained from organic materials from meteorites. The packages sent to Mars successfully collected samples and subjected them to the analytical procedures but, to the great surprise of the investigators, no organic molecules were detected. This does not necessarily indicate a complete absence of carbon compounds, as the sensitivity of the instruments was not great enough to detect amounts less than the equivalent of the compounds present in Earth soil containing about one million bacteria per gram. This result, while interesting chemically, was not conclusive biologically, as it was certainly possible that Martian organisms, if present, might not be plentiful. Other possibilities were raised, for instance that the organisms might have hard coverings trapping the organic materials or might scavenge all available nutrients, so keeping the soil content low. In addition, organic molecules exposed at or near the surface might be destroyed by the intense solar radiation or by reacting with highly oxidized surface materials.

The gas-exchange experiment

This involved the collection of a sample of Martian surface material and mixing it with a solution of substances (called a *medium*) known to be necessary to support the growth of various Earth micro-organisms. Utilization of the nutrients should give rise to changes in the composition of the gas mixture above the medium, and these can be measured by mass spectrometry. The design of the experiment was based on the assumption that micro-organisms in some ways similar to those on Earth might have evolved on Mars, and that there was a reasonable possibility that Martian and Earth organisms would share some metabolic capabilities. A negative result would not in itself indicate absence of Martian life, as organisms with different nutritional requirements might be present. It was also possible that some constituents of the complicated mixture provided might be poisonous for Martian organisms. Very slowly metabolizing organisms would also be missed. False positive results were a possibility, too, because of chemical reactions of unanticipated kinds with constituents of the Martian regolith. Physical release of gases held in porous material is another possibility. As biological reactions, at least on Earth, are in general eliminated or greatly reduced by heating the sample before testing to 145°C, it was arranged to include this as part of the experiment in the hope of distinguishing between biological and non-biological reactions. It was considered desirable to remove the medium from time to time and replace it with a fresh solution. If a biological growth process were taking place, there should be new episodes of gas exchange with each fresh batch of medium.

In the actual experiment, the Martian regolith material was first exposed to water vapour. This in itself might have encouraged growth of Martian organisms if nutrient materials were present in the Martian 'soil'. With samples collected at both sites, exposure to water vapour led, totally unexpectedly, to the release of oxygen, together with carbon dioxide and nitrogen. By careful manipulation of the collection equipment, controlled from Earth by radio signals, a rock was moved aside to permit collection of a sample that had not been exposed to sunlight, as it was considered possible that Martian organisms in the surface material might survive better in a protected environment. The result, however, was the same as with exposed regolith. Heating of samples to 145°C before humidification produced discrepant results, as at one site the oxygen release was abolished, whereas at the other it was cut to half, but there may have been some fault in the functioning of the equipment. Carbon dioxide release was abolished but nitrogen release was not significantly affected.

Gas changes were observed when the solutions of nutrients were added to the samples, but the results were difficult to interpret as they did not fit any of the patterns observed with Earth organisms.

It is perhaps not surprising that, in the virtually unknown Martian chemical environment, an experiment based on terrestrial experience gave confusing results. With luck, it could have been otherwise, but in spite of the superb technical achievement of the mission, this experiment left the question of Martian life open. Chemists have not been able to provide a completely convincing alternative to a possible biological explanation, but that too is understandable in view of the unfamiliar nature of Martian surface chemistry. It has been suggested that compounds known as superoxides might be present on Mars and produce at least some of the observed responses, and oxides of manganese may also be a factor. Most investigators favour non-biological chemical explanations for the results of this experiment.

The labelled release experiment

When organisms process nutrients in their cells, some of the carbon present in the molecules is ultimately released as carbon dioxide (CO_2). If some of the carbon in the nutrient molecules is present as a radioactive isotope (C^{14}), then some of the CO_2 released is likely to be radioactive, and this radioactivity can be detected in gases above the growth medium. The release of radioactive carbon dioxide as an indicator of metabolism and growth is now commonly measured in terrestrial laboratories, but the idea of using this approach seems to have originated in the early work on Martian probes.

If Martian organisms were able to use nutrients similar to those required by Earth organisms, the same detection methods for growth and metabolism should be applicable. As with the gas-exchange experiment, possibilities for false negative and positive results existed. To eliminate biological activity, preheating of some samples to 160°C was included, and again there was provision for successive removal and addition of medium to see if a continuing activity could be detected, as would be expected for biology but probably not for a purely chemical explanation.

The results of this experiment on Mars appeared to be positive, as samples collected at both sites gave release of radioactive carbon dioxide. There was also evidence that phases of uptake of carbon dioxide by something in the sample could occur, as seemed to be the case when fresh medium was added. Heating of the soil to 160°C abolished ability of the sample to release carbon dioxide, and even heating to 50°C led to a great reduction in release.

Again, unfortunately, the results are tantalizing rather than conclusive, and might be attributed to either exotic chemistry or to biology. Some reviewers of the data have declared in favour of a biological explanation, but to others, particularly because of the absence of direct evidence for the presence of organic materials in the samples analysed, and the growing evidence that Martian surface chemistry is strange and unfamiliar, a non-biological explanation seems to be indicated. Chemical explanations are possible, although the presence of the necessary compounds on Mars has not been directly demonstrated. The pro-biology lobby has been particularly impressed by the marked inactivation produced by heating to the lower temperature of 50°C, and have maintained that there is an unreasonable reluctance to face the seemingly positive evidence provided by this experiment. It would appear, however, that in a matter so important, a prudent scientist would reserve judgement at this stage, for it is evident that Mars has faced chemists, as well as biologists, with new problems on which there is little basic information available. The enlargement of chemical horizons is undoubtedly one of the valuable outcomes of the Mars missions.

The carbon-assimilation experiment

(Note: sometimes referred to as the *pyrolytic release* experiment.) In an earlier chapter it was explained that organisms able to carry out photosynthesis make use of carbon dioxide for building up more complex molecules. Non-photosynthesizing organisms may also fix some carbon dioxide from the atmosphere but to a lesser extent. Carbon monoxide was included, too, as some micro-organisms are able to make use of it. Both the uptake processes are detectable in tests carried out on moist terrestrial soils containing active organisms, and an experiment was designed to be carried out on Mars. In this, Martian conditions were simulated inside the apparatus, including the provision of simulated Martian sunlight. As laboratory experiments had shown that ultra-violet radiation could promote the synthesis of organic compounds in a Martian-type atmosphere, the light in the Martian experiment was filtered to remove the active wavelengths. The soil samples were moistened for some tests but not for others. With Earth organisms, absence of moisture caused the test to be negative. It was considered that the design of this experiment minimized the risk of a false positive result, but as with the other experiments, the possibility of Martian organisms having different requirements from those on Earth meant that a negative result would not rule out the presence of life.

Tests carried out at both sites gave weakly positive results, but neither light nor moisture seemed to make a significant difference to what was found. Soil that had been heated to 175°C for three hours before being exposed to the gases showed a very low activity, but heating to 90°C for two hours had only a slight effect.

These results are extremely difficult to interpret, as too little is known about the chemistry of Martian regolith to permit confident assertions to be made. The failure of heating to have as great a destructive effect as would have been found with Earth organisms does not in itself mean that the observed assimilation of carbon was non-biological. Nevertheless, because of the findings in the other experiments and the growing suspicion that Martian surface materials may be in a state with which terrestrial chemists are unfamiliar, most investigators have come down in favour of a non-biological explanation. Laboratory simulation of the Martian results by non-biological processes has been to some extent achieved. A final decision could only be reached after further Martian exploration and experimentation, but in the meantime the outlook for Martian biology seems, to say the least, to be most unpromising.

The Viking experiments have brought into focus some of the difficulties faced by those searching for extraterrestrial life. When experiments are carried out on Earth, we are often sufficiently familiar with the environment to be able to rule out some processes with reasonable certainty. In an alien environment, largely unknown and not studied in detail by chemists, alternative explanations to those we would favour if the experiments were being carried out on Earth are more difficult to eliminate, and problems encountered are not necessarily easy to foresee. This is why visual revelation of the presence of obvious organisms would have been easier to accept as evidence, and even now a few hopefuls cling to the supposedly changing greenish patches.

In the 1950s and 1960s, several attempts were made to test the ability of Earth-type organisms to survive in a simulated Martian environment. It was shown that the tolerance of organisms for extreme conditions was greater than many had supposed, but the information on the parameters of the Martian environment was sadly defective, and now we would suspect that the Martian surface materials, at least in the sites examined so far, would be highly unfavourable to Earth organisms, even to the point of being swiftly destructive. In the last few years the pro-lifers have drawn some hope from the discovery in Antarctica of lichens (mutually dependent fungi and algæ) and bacteria living in porous rock, an environment which provides some protection from the extreme antarctic conditions. The studies of Friedman and Ocampo have shown that a black fungus grows just below the surface of the rock, and filters out some of the intense summer light. Deeper, but less than half an inch from the surface, is a colourless fungus, and immediately below this a band of algæ, which are green. These organisms make use of the minerals of the rock, the carbon dioxide of the air and obtain water from snow. Interestingly, it is suggested that the auroral activity leads to the production of nitrogen compounds which are carried to the surface by snow. To quote Friedmann (1983), 'What the algæ produce by photosynthesis, the fungi consume; the bacteria clean up by feeding on dead algæ and fungi. Winter's freeze simply puts the lichen into suspended animation until the summer thaw.' Friedmann has commented that if, when Mars had its water in liquid form, life did in fact develop there, Martian organisms could have adopted a mode of life similar to that found in Antarctica. At this stage, further speculation is unlikely to be fruitful, but should another mission be sent to Mars, investigation of this remote possibility should be undertaken. It would be desirable, too, to place probes near a polar cap, where more water is likely to be available to any organisms that might be present. Our feeling, however, is that Mars is probably a sterile world, and has always been so.

Future colonization of Mars

Despite its disadvantages, there can be little doubt that Mars will be the first planet to be reached by a human expedition, and plans for a Martian base are already being drawn up. It is an open secret that the Russians, too, are interested in the possibilities. Water will almost certainly be available, and in this respect Mars is much more accommodating than the Moon.

The pioneer expedition will have to establish a base, since a period of months after arrival must elapse before the relative positions of the Earth and Mars will permit a return journey to be started. What form the base will take remains to be seen, but the old science-fiction idea of a pressurized dome may not be so very wide of the mark. As on the Moon, pressurized suits will be needed in the open, and powered exploration vehicles could be constructed. Information on the physiological effects of prolonged space flight and exposure to 'weightlessness' is being obtained by the Russians, whose cosmonauts have spent long periods in Earth-orbiting space stations. Loss of calcium with weakening of the bones is a potential problem. It is likely, too, that severe psychological problems arising from isolation might be encountered. Selection and training of personnel will have to be rigorous. Experiments in the USA on prolonged isolation of a few individuals in a simulated Martian artificial 'biosphere' should help to solve some of the problems but even a simulated situation on Earth will not have the psychological impact of the real thing. The ethical consequences of Martian exploration, including the fate of children who might be born there, must also be faced.

These are problems for the future, but Mars extends a challenge which humanity surely cannot ignore.

VI VENUS

Before the Space Age, it was widely believed that as an abode of life Venus might be just as suitable as Mars – perhaps more so. Unfortunately for the biological optimists, this has proved not to be the case, and Venus is now known to be extremely hostile to terrestrial-type life, even though in size, mass and escape velocity it is almost a twin of the Earth. Essential data are as follows:

Distance from the Sun	Max. 109.0 million km.
	Min. 107.4 million km.
	Mean 108.2 million km.
Sidereal period	224.701 days
Orbital eccentricity	0.00678
Orbital inclination	3°.394
Axial rotation period	243.01 days
Axial inclination	178°
Diameter	12,102.8 km.
Mass (Earth = 1)	0.815
Volume (Earth = 1)	0.86
Density (Water = 1)	5.24
Escape velocity	10.36 km/sec.
Surface gravity (Earth = 1)	0.902

The 'solar day' on Venus is 117 Earth-days, i.e. a 'daylight' period of 58.5 days followed by a 'night' of equal length. The rotational axis is inclined by only 2 degrees to the perpendicular to the orbit, but the rotation is retrograde – for reasons which are still quite unknown. Venus has no satellite.

Surface markings

From Earth we can never see the true surface of Venus, which is permanently concealed by the dense, cloud-laden atmosphere, and therefore attempts at mapping by telescopic means have been doomed to failure, despite many attempts (Moore, 1985). Often the disc appears blank; at other times vague, impermanent features may be seen, but are always difficult to fix with any accuracy. Before the flight of Mariner 2, in 1962, very little was known about Venus, which was often described as 'the planet of mystery'. Even the length of its rotation period was unknown, and theories of the surface conditions ranged from a raging dust-desert to a planet-wide ocean! A summary of historical work, together with modern results, has been given by Hunt and Moore (1982); more technical details are contained in a major book edited by Hunten (1983).

Rotation period

Before 1962, many attempts were made to determine the rotation period, using visual, photographic and theoretical methods. Moore (see Hunt and Moore, 1982, p. 167) listed over eighty of them, the results ranging from 23 hours up to 224.7 days – every one of which proved to be wrong. The suggested 224.7-day period is, of course, the same as that of the Cytherean* year, and would mean that the rotation would be synchronous, so that the planet would keep the same face turned permanently sunward; there would be only slight libration effects, since the orbit is so nearly circular. However, from 1962, it has become clear that the true period is even longer than this – a case unique in the Solar System, at least so far as we know. In theory, an observer on Venus would see the Sun rise in the west and set in the east, though in practice the clouds would prevent the Sun from being visible at any time.

Atmosphere

In any study of Venus, the composition of the atmosphere is obviously of paramount importance. As early as 1932 Adams and Dunham, using spectroscopic equipment on the 254-cm. (100-inch) Hooker reflector at Mount Wilson, detected infra-red absorptions that were considered to indicate the presence of carbon dioxide, and this has proved to be correct. Since carbon dioxide is an important factor in the production of the so-called 'greenhouse effect', which effectively traps solar heat, it was probable that the surface temperature would be very high. Modern results have confirmed that the atmosphere is mainly carbon dioxide, with relatively small amounts of other gases, including argon. The atmospheric pressure at ground level is at least ninety times that of the Earth's air at sea level.

*There is still no universally accepted adjective for Venus. The common *Venerean* and *Venusian* are ugly; we prefer *Cytherean,* from an old Sicilian name for the planet.

In 1937, following a visual and photographic study, F. Wildt proposed that the clouds were likely to be made up of formaldehyde (H.CHO), formed under the influence of solar ultra-violet radiation; later, in 1955, F. L. Whipple and D. H. Menzel suggested that the main constituent was more likely to be ordinary water. Both these ideas have been disproved by the space-probe results, and we now have an excellent picture of the cloud structure and nature (see Krallenberg *et al.*, 1980). The clouds contain a great deal of sulphuric acid, a fact which would have important implications for theories of a possible atmospheric biosphere.

Apparently the top of the atmosphere lies about 400 kilometres above the surface, and the upper clouds lie at 70 kilometres. At an altitude of 50 kilometres the temperature is about 20°C; below lies a clear layer, then a layer of denser cloud. Beneath this, at 47 kilometres, there is a second clear region, and the cloud-deck ends at 30 kilometres above the surface, so that the lowest part of the atmosphere may be aptly described as corrosive, superheated smog. The surface temperature is of the order of 465°C.

The wind patterns are remarkable, since the upper clouds have a rotation period of only 4 days. Winds decrease close to the surface, from about 100 metres/sec. at the cloud-top level to only 50 metres/sec. at 50 kilometres, and only a few metres/sec. at the surface; though in such a dense atmosphere, even a slow wind will have tremendous force.

The surface conditions: older ideas

In the pre-Mariner period, theories about the surface conditions were many and varied. Mention has already been made of the 'Carboniferous' picture proposed by Svante Arrhenius (1918), who maintained that much of Venus was covered by swamps:

> The temperature is not so high as to preclude a luxuriant vegetation. . . . The vegetative processes are greatly accelerated by the high temperature. Therefore, the lifetime of organisms is probably short. . . . Later, the temperature will sink, the dense clouds and the gloom disperse, and some time, perhaps not before life on Earth has reverted to its simpler forms or even become extinct, a flora and fauna will appear.

Arrhenius was not alone in his views; as recently as 1954 N. A. Kozyrev estimated a surface temperature of only 30°C. In 1955, Sir Fred Hoyle proposed that there might be oceans not of water but of oil, so that 'Venus is probably endowed beyond the dreams of the richest Texas oil-king.' Menzel and Whipple (1955), who believed the clouds to be of water, suggested that the surface might be almost entirely oceanic. This led to a rather bizarre corollary. If there were seas, then the water would have dissolved much carbon dioxide, producing oceans of soda water. No-one, however, suggested volcanoes spewing out whisky.

If any of these ideas had been correct, it was reasonable to suppose that primitive life, at least, might have developed on Venus, in a manner similar to its development on Earth. However, the results from Mariner 2 showed that the surface is intolerably hot for any type of carbon-based life. The general picture suggested that it was most unlikely that Cytherean life existed.

The surface conditions: current facts and theories

Mariner 2 was followed by other probes, American and Russian. The first probe to land on the planet was the Soviet Venera 3, in 1966. It failed to return any data, as it was crushed by intense pressure as it parachuted through the atmosphere. Four years later Venera 7 made a successful soft landing, surviving to transmit data for twenty-three minutes. Then, in 1975, Veneras 9 and 10 sent back pictures directly from the surface, showing a rock-strewn, desolate and dimly-lit terrain. Subsequently, the American and Russian orbiters, beginning with Pioneer Venus in 1978, have sent back detailed radar maps of almost the entire surface, and have confirmed the extreme temperatures and pressures, while in 1985 the Soviet Vega probes, en route to a rendezvous with Halley's Comet, dropped balloons into the Cytherean atmosphere and obtained information from various levels.

The maps indicate that there is a huge, rolling plain which covers some 60 per cent of the surface; there are rift valleys, one of which, *Diana Chasma*, has its lowest point 2 kilometres below the mean radius; and there are two main upland areas, *Ishtar Terra* in the north, with a diameter of 2900 kilometres, and the equatorial *Aphrodite Terra*, which is larger (9700 × 3200 kilometres). The *Maxwell Montes*, occupying the eastern end of Ishtar, rise to 11 kilometres above the mean level of the planet. There are also volcanoes which are probably active, notably *Rhea Mons* and *Theia Mons* in the smaller upland area of *Beta Regio*. Venus, unlike the Earth, seems to be a 'one-plate' planet, so that Earth-type plate tectonics do not apply, and the volcanic regions are localized.

Bearing these facts in mind, what are the chances of Cytherean life, past or present?

Whether there are any niches on or within Venus in which carbon-based life could exist will be hard to prove, but it does seem most improbable. The planet we now observe has been drastically modified by a runaway greenhouse effect. Only about 1 per cent of the sunlight reaching the top of the atmosphere penetrates to the surface, but the carbon dioxide absorbs heat radiated from the surface in the infra-red part of the spectrum and re-radiates some of it so that surface heating can occur. The surface appears to be almost certainly lifeless.

Major speculations about the possibility of life on Venus have in recent years shifted from the surface to the atmosphere. It has been suggested that, before a runaway greenhouse condition developed, there may have been enough water and a suitable temperature to permit prebiological and primitive biological evolution to proceed to some extent. It has been proposed that life could have shifted to, or have evolved in, the atmosphere (an atmospheric origin of life has been postulated for the Earth, too – see Woese, 1980). Organisms might therefore float in the atmosphere at levels where conditions are more tolerable than at the surface. The presence of the sulphuric acid clouds would not necessarily be incompatible with the existence of some forms of life. To us it seems that the chance even of atmospheric carbon-based Cytherean life is vanishingly small. 'Inorganic life' has been postulated in the form of a planetary organism, a Cytherean counterpart of Gaia, but quite different in constitution, perhaps with germanium replacing carbon and energy derived from electrical cells using sulphuric acid as an electrolyte. Moreover, it has been suggested that such an 'organism', in effect a sort of 'planet-wide computer', might even have arranged the rapid rotation of the atmosphere to even out temperature over the surface (Chalmers referred to by Lunan, 1979). The ingenuity of such fascinating speculations is probably not matched by the probability of their truth, but they should not be decried, especially when they are or soon will be susceptible to testing.

It is probable that when Venus and the Earth were formed, the impact of planetesimals produced heating, and that in the course of time after cessation of impacts a condition of equilibrium was reached in which the rate of loss of radiation in the infra-red was balanced by incoming visible and ultra-violet solar radiation. Some astronomers now believe that Venus never became cool enough for rain to occur. Water existing as vapour in the atmosphere would have been dissociated by light to give hydrogen, a light gas which would have been lost to space, and oxygen which has ended up in combination as carbon dioxide.

If, even in the early days of formation of the Solar System, when the Sun was up to 30 per cent less luminous than it is now, Venus was nevertheless hot because of planetesimal impacts, and never had significant amounts of liquid water at its surface, the prospects for past life seem to be dim. An alternative possibility is that oceans did exist for a while but boiled away as the intensity of sunlight increased, contributing to the development of the runaway greenhouse effect. Any primitive organisms that had developed would have been destroyed. Even analysis of the planet's surface, in its present condition, may not help in assessment of the probability of past Cytherean life.

Terraforming Venus?

There seems to be no chance of any manned expeditions to Venus, at least in the foreseeable future. A suggestion, due initially to Carl Sagan, that it may be possible to 'seed' the clouds, releasing free oxygen, cannot be ruled out, but it is beyond present technological capabilities. For the moment, we must be content to examine the planet from a respectful distance, and in our search for life we must look elsewhere.

VII OTHER WORLDS OF THE SOLAR SYSTEM

Conditions on Mars and Venus may not be promising for terrestrial-type life, but other bodies of the Solar System are equally unsuitable or even less suitable. Most of them lack appreciable atmospheres, the only exceptions being the giant planets and Saturn's satellite, Titan, so that we must first look at the possibilities – if any! – of life upon an airless world.

An atmosphere might indeed play a decisive rôle in the development of life, especially if biopœsis occurs at the surface or in oceans. The effects of an atmosphere are both physical and chemical, as discussed elsewhere in this book, and atmospheric pressure will determine what liquids can exist at the planetary surface, and will determine boiling points of liquids. Organisms make use of atmospheres as sources of materials and for getting rid of certain waste products. Atmospheres exert filtering effects on radiations, and may, by the greenhouse effect, influence surface temperature.

We may be reasonably sure that a completely dry, atmosphereless world would not give rise to life – our own Moon is an example. Organisms formed on a world that was losing its atmosphere but had supplies of water might penetrate to underground situations and live in the depths, depending on energy derived from within the planet. Whether tough, impermeable portions of an organism exposed at the surface of an airless world could make use of sunlight and contribute to subterranean life is a matter for speculation. Probably, on such a planet, a population of hardy micro-organisms would have the best chance of survival, followed by any intelligent beings who might have developed an advanced technology.

Mercury

We need say little about Mercury, the innermost planet, whose distance from the Sun ranges between 69,700,000 and 455,900,000 kilometres. The diameter is 4880 kilometres (smaller than that of Jupiter's satellite Ganymede) and though the planet is dense, with a specific gravity of 5.5, the escape velocity is only 4.3 kilometres/sec. The sidereal period is 88 days, and the rotation period 58.6 days – two-thirds of a Mercurian year.

So far only one space-probe has by-passed Mercury; this was Mariner 10, which made three active rendezvous with the planet (March and September 1974, and March 1975), sending back excellent pictures of wide areas of the surface. Previously, the best map was considered to be that by E. M. Antoniadi (1934). Antoniadi believed that the rotation was synchronous, and the atmosphere dense enough to support clouds; but both these conclusions have been disproved, and Antoniadi's map, not surprisingly, bore little relation to the truth. From the Earth, Mercury appears very small and is excessively difficult to study in any detail.

The Mariner 10 pictures showed that the surface is mountainous and cratered, with one vast basin, *Caloris*, with a diameter of 1300 kilometres. Though the features are not identical in nature with those of the Moon, there is an obvious similarity. The atmospheric pressure is no more than 10^{-9} millibars, and the main constituent is helium, suggesting that the weak magnetic field traps helium nuclei from the solar wind. Full details were given in a special edition of *Science* (1974). The mean surface temperature is 350°C (day) and −170°C (night).

The extreme temperature range at the surface, the lack of water and of a significant atmosphere would seem to rule out indigenous carbon-based life, and it is likely that Mercury never possessed an atmosphere of a type which would favour biopœsis. Little is known about the interior of Mercury, but it has been suggested that it may be somewhat Earth-like. It is almost certainly sterile. Mineral organisms and 'physical life' remain possibilities (highly speculative) in an environment of this kind.

Pluto

In the far reaches of the Solar System we find Pluto, discovered in 1930 by Clyde Tombaugh at the Lowell Observatory in Arizona. The orbit is very eccentric; the distance from the Sun ranges between 7375 million and 4425 million kilometres – so that near perihelion Pluto is closer-in than Neptune. Full details have been given by Tombaugh and Moore (1980). The revolution period is 248 years, and the rotation period 6 days 9 hours.

The diameter of Pluto is only 3000 kilometres, and the density is low. The surface is probably covered with methane frost. The satellite or companion body, *Charon*, was discovered in 1977, and has perhaps one-third the diameter of Pluto itself.

Pluto's mean temperature is of the order of −230°C. There is probably a very thin methane atmosphere. We can rule out Pluto as a site of active, carbon-based life.

The giant planets

For most practical purposes we may consider the four giants together, though they are by no means exactly similar, and the Jupiter/Saturn pair differs markedly from the Uranus/Neptune pair. Jupiter has been surveyed by four probes – Pioneer 10, 1973; Pioneer 11, 1974; Voyagers 1 and 2, 1977 – while Saturn was briefly surveyed by Pioneer 11 in 1979 and by Voyagers 1 (1980) and 2 (1981). Voyager 2 then by-passed Uranus (1986) and is currently on its way to a rendezvous with Neptune in 1989. (For details, see Kondratyev and Hunt, 1982; Morrison, 1982; Hunt and Moore, 1987.)

As we have seen, the giants have relatively small cores, surrounded by liquid layers (mainly hydrogen) which are overlaid by gaseous atmospheres, in which hydrogen and helium are dominant. Jupiter, Saturn and Uranus have strong radiation zones surrounding them, particularly strong in the case of Jupiter, and there are powerful magnetic fields; the same is probably true of Neptune. Obviously there is no hope of manned expeditions to these planets at any time.

Jupiter

The chances of finding carbon-based life on Jupiter seems to be exceedingly remote, in spite of the ingenuity of some speculative authors (e.g. Sagan and Salpeter, see Baugher, 1985) in proposing atmospheric organisms they have termed 'sinkers', 'floaters' and 'hunters'. Sinkers are proposed as products of the upper atmosphere, and to feed on organic molecules of the clouds. Ultimately, they sink and are 'eaten' by 'floaters', organisms supposed to inhabit the deeper levels of the atmosphere. The third kind of organism, the 'hunter', is visualized as being several kilometres in diameter, balloon-like and able to move, perhaps by some sort of jet propulsion, through the atmosphere. Hunters eat floaters.

Unfortunately, there are reasons to believe that the Jovian atmosphere is more likely to be of purely chemical than of biological interest. Current estimates suggest that water may not be readily available. Haldane (1954) pointed out that there is a whole system of inorganic and organic chemistry in which liquid ammonia takes the place of water. This raises the possibility that at some level in the Jovian sphere low-temperature evolution might have led to the production of ammonia-based systems analogous to the water-carbon organisms we know. This has caught the imagination of many authors, but we consider it to be unlikely (see Ch. X).

The atmosphere is probably extremely turbulent, and complex structures would probably be destroyed by being circulated to the hotter deep regions of the atmosphere. The composition of the coloured components of the atmosphere is unknown, but possibly organic molecules may be present, as suggested by Briggs (1960). The atmosphere, mainly hydrogen and helium, does contain some methane, ammonia and enough water to permit synthesis of organic molecules. Lightning has been photographed and could be a source of energy for syntheses. This large planet's star-like qualities produce a considerable energy flux, and the turbulence of the atmosphere would tend to mix components from various levels, with persistence of more complex molecules in the upper, cooler regions.

It is certainly possible that Jupiter might hold many surprises for future investigators. The great size of the planet, its variety of zones with different conditions, varying from very hot in the depths to extremely cold in the outer regions, the range of pressures, the strong gravitational field and the magnetosphere together constitute a system of great complexity. Lederberg (1960) commented that the abundance of light elements in the Jovian atmosphere, subject to solar radiation at low temperatures in a high gravitational field 'offers the most exciting prospects for novel biochemical systems'. Only future investigation will provide the data to test the hypotheses that have been advanced.

Saturn

Saturn is smaller than Jupiter and much farther from the Sun. The atmosphere is probably drier than Jupiter's and the feebler colours may indicate that organic molecules are not as readily synthesized as in the Jovian atmosphere. Saturn is almost certainly not a seat of carbon-based life.

Uranus

Unlike the other giants, Uranus seems to have no appreciable internal heat-source. The magnetic axis is inclined to the rotational axis by 60° – a case unique in the Solar System – and the rotational axis is inclined at 98° to the perpendicular, which means that the rotation is technically retrograde. The reason for this is unknown. There is more water and methane than on Jupiter. It is highly improbable that active carbon-based life exists at any level in Uranus.

Neptune

Neptune does have an internal heat-source, and the rotational axis is not tilted to the same extent as that of Uranus. At the moment our knowledge of the planet is limited, and we must pin our hopes on the success of the Voyager 2 rendezvous in 1989. No more probes to the outer Solar System have been planned as yet, apart from the Galileo mission to Jupiter, which has been delayed because of the 1986 Challenger Space Shuttle disaster; whether the Russians will attempt any such missions remains to be seen.

Planetary satellites

We must now consider the satellites of the planets. The Moon has already been discussed. Of the rest, only the Galilean satellites of Jupiter, Saturn's satellite *Titan* and Neptune's satellite *Triton* are of planetary dimensions. The rest are so small, and so lacking in mass, that they can never have had appreciable atmospheres or supplies of liquid water, and they will not be considered here.

The Jovian Galilean satellites

The four major attendants of Jupiter have been surveyed in detail by the two Voyager probes. Analysis of the data collected has led to proposals for models of the satellites, and has aroused some interest among would-be exobiologists. Europa and Ganymede have been the two of greatest interest as potential sites of biological activity, despite their lack of appreciable atmospheres. It must, however, be stressed that in spite of the remarkable achievements of the Voyagers, the present models are necessarily tentative and subject to later revision.

The Galileans lie within the Jovian magnetosphere, and are subjected to bombardment by high-energy particles. Jupiter is very large, with a hot core, and it has been estimated that after its formation there would have been a period of about 100 million years during which ice and water could have been retained at the distances of Ganymede and Callisto, whereas closer to the planet, in the regions of Io and Europa, energy released from Jupiter would have rendered the existence of ice and water unlikely. Instead, water-enriched silicates would have been present.

Io

This little world, slightly larger than our Moon, has caused great excitement because it is the site of active vulcanism. Its surface appears orange, with yellowish-white blotches, and shows no evidence of impact-craters. Dark volcanic vents are scattered over the surface, and active volcanoes have been observed. The core of Io is believed to consist of rock and metals, and the outer crust contains sodium and potassium compounds. It seems likely that there are sub-surface flows of molten sulphur, and collections of liquid sulphur dioxide (SO_2). Hot sulphur from the interior may cause vaporization of the sulphur dioxide which emerges from vents at the surface and solidifies to form a snow-like deposit. Molten sulphur also flows from the vents and, after cooling, forms a yellow surface deposit.

Io can, with virtual certainty, be ruled out as a site of carbon-based life. The unlikelihood of the existence of sulphur-based organisms is mentioned elsewhere (Ch. X). Only the remote speculative possibility of 'physical life' remains.

Europa

Computer-enhanced pictures of the surface of Europa show a network of fine lines and occasional impact craters, but nothing resembling the volcanic vents of Io. The absence of large numbers of impact craters has been interpreted as evidence that the surface is sufficiently plastic to lead to disappearance of craters in a short time, possibly because there is an icy crust kept 'warm' by thermal energy from within the satellite. The strange lined appearance of the surface has been likened to that of expanses of polar ice on Earth, but there is no completely satisfactory explanation of the features observed. There is evidence that Europa is rich in water, metals and silicates. It was postulated that there could be a vast ocean of water about 100 kilometres deep under an ice crust about 40 kilometres thick, but it is perhaps more likely that most of the water is in the form of an ice layer. The temperature at the surface ($-140°C$) would appear to rule out active carbon-based life, and the satellite has no significant atmosphere.

It has been suggested that sunlight might penetrate to the interior around cracks in the ice, and that zones as hospitable to carbon-based life as some surface-frozen but biologically active lakes in the terrestrial Antarctic might exist in a liquid layer near the cracks. It is therefore remotely possible that Europa might support life, but the chance that carbon-based organisms arose on Europa by any of the commonly postulated mechanisms appears to be very small. The possibility of prebiotic syntheses and even the production of organisms in the depths of an ocean in areas heated from the interior, in some but not all ways resembling the terrestrial 'black smokers', cannot be completely ruled out, but the correct starting materials would have had to be available. The energy for the syntheses would have been derived largely from the interior, and the chance of sunlight playing a significant role in the depths of a sub-ice liquid layer would seem to be virtually zero. If a significant liquid layer persisted for only a short time, we may suspect that prebiotic syntheses, if any, might not have progressed far.

Ganymede

Analysis of data collected by the Voyager probes and by other means has led to the suggestion that silicates and water are constituents of this large satellite, which has no significant atmosphere. The surface is apparently thick ice (50 km or more) which has been cratered by meteor impacts. The surface features point to tectonic activity, and there is an extensive groove system which might have been caused by tensions in the surface layers. It is considered possible that beneath the ice is an ocean of water whose volume is perhaps much greater than that of the Earth's water, but to what extent it is liquid or an ice/water mixture is uncertain. It is also possible that it is mostly ice – an icy mantle under an ice crust. It has been proposed that although the surface temperature of Ganymede's ice would be low (−170°C, indicated by measurements), the depths of an ocean beneath the ice would be warmed by heat from the core, and this could give a zone in which there are temperatures suitable for carbon-based life. No sunlight would penetrate to the postulated sub-surface ocean, so it would not be available for photosynthesis. The energy for prebiotic syntheses and for maintaining life would have to be derived ultimately from radioactive decay in the core, and appropriate starting materials would have to have been available. The remoteness of Ganymede from the Earth and the protection of the supposed ocean by enormously thick ice would seem to present major difficulties for attempts at exploration by automated devices.

Callisto

The structure of this heavily cratered satellite is probably similar to that of Ganymede, and a very thick surface ice layer would blot out all solar radiation. The situation for biopœsis is equally unpromising.

Saturn's Moon, Titan

Titan has a dense atmosphere, perhaps about 1.5 times as dense as that of the Earth. The major constituent of the atmosphere is nitrogen, which accounts for 90 per cent or more of the molecules present, the rest being mainly methane. The satellite apparently consists of a rocky core covered by a thick ice layer, and there is a possibility that a layer of liquid water exists beneath the ice. The frozen water is too cold to sublime into the atmosphere to a significant extent and is consequently holding its oxygen, so that it could not be released from the water molecules by photolysis in the atmosphere or contribute to other molecules by water taking part in atmospheric reactions. The atmosphere is probably the site of formation of organic compounds, including nitrogen-containing compounds produced by reaction of hydrogen cyanide with other constituents such as dissociation products of methane. In fact, there seems to be a virtual smog in the atmosphere, especially at high levels. Organic compounds would settle slowly to the surface and remain there because of the low temperature. Vast quantities of hydrocarbons might therefore cover the surface of Titan, and it has been suggested that oceans of liquid methane mixed with other hydrocarbons might be present.

At this time it is possible only to say that Titan is another site in the Solar System worthy of further study to see if it might shed light on prebiotic chemistry, or more probably on the abiogenic production of hydrocarbons. Direct exploration of Titan by automated probes is a possible task for the distant future.

Triton

Triton, the large satellite of Neptune, is unusual inasmuch as it has retrograde motion; all the other retrograde satellites in the Solar System are very small, and probably captured asteroids. Triton, discovered by W. Lassell in 1846 (only a few weeks after the identification of Neptune itself, by Galle and d'Arrest from Berlin), is certainly large; it is probably larger than our Moon, and is therefore far superior to any of the satellites of Uranus. Its mean distance from Neptune is 333,000 kilometres, and its orbit is almost perfectly circular.

Indications of a very tenuous atmosphere have been reported, but the temperature is very low (around $-220°C$) and the chances of life there are probably negligible. However, we must admit that at present our knowledge is fragmentary, to put it mildly, and once again we look forward to the success of Voyager 2 in 1989. Triton, like some other satellites, may prove to be a surprising body.

Summary

On the whole, leaving aside the Earth, the search for carbon-based life in the Solar System has so far been unsuccessful, or at best inconclusive. The grosser beings of imagination, such as Lowell's canal-building Martians and Arrhenius' amphibia, have been consigned to the realm of myth. Intelligent life seems to be ruled out. The possibility that micro-organisms might exist elsewhere has not been thoroughly investigated, but the experiments on Mars seem to point more to chemistry than to biology. The situation, size and constitution of the Earth may indeed have special features that make it the only possible abode for the production, or persistence, of life in the Solar System. Nevertheless, direct investigation has barely scratched the surfaces of other worlds, and future research may reveal unsuspected facts, but our personal view on the chances of finding extraterrestrial carbon-based life in the Solar System is now pessimistic.

VIII COMETS AND METEORITES

It is quite wrong to suppose that the Solar System consists only of the Sun, the planets and their satellites and asteroids. As we have seen, there are also vast numbers of comets and meteoroids, together with interplanetary 'dust'. Although comets and meteorites may not at first sight be regarded as possible sites for living organisms, some interesting proposals have recently been made. We have mentioned earlier the theory of *lithopanspermia*, and the presence of organic chemicals in certain meteorites.

Comets

Comets are essentially icy. When a comet nears the Sun, the ices begin to evaporate, producing a coma and tail or tails. It is thought that the comets are made up of very primitive material, and that there is a whole cloud of them orbiting the Sun at a distance of more than a light-year; this is called the *Oort Cloud*, after the Dutch astronomer, Jan Oort, who first suggested its existence, and it may contain more than a hundred billion comets. If a comet is perturbed for any reason, it starts to fall toward the Sun. It may then simply swing through perihelion (its closest position to the Sun) and move back to the cloud, not to return for an immense period; or it may be perturbed by the gravitational pull of a planet (usually Jupiter) and be forced into a short-period orbit. Since a comet loses material every time it passes through perihelion, it is bound to be short-lived on the cosmic scale, and indeed several periodical comets seen regularly in past years have now disintegrated. The only periodical comet which may be bright is Halley's, which has a period of 76 years and was last at perihelion in 1986 – though on that occasion it was a disappointing spectacle because it was so badly placed.

Five space-probes (two Japanese, two Russian and one European) were sent to rendezvous with Halley's Comet in March 1986. The European spacecraft, *Giotto*, actually penetrated the coma and sent back close-range pictures of the nucleus, which, contrary to the expectations of most astronomers, proved to be very dark (reflectivity about 2 per cent) rather than bright, so that it seemed possible that the icy core was overlaid with blackish material.

Proposals of Hoyle and Wickramasinghe

Two astronomers who were not surprised by the dark nucleus were Sir Fred Hoyle and his colleague Chandra Wickramasinghe, as they had predicted that it would be so. Hoyle (1986) has recently argued that the black surface is not due to absorption of sunlight at the surface, as the surface is too cool for this to be the explanation. Rather, he suggests that the surface materials are translucent to optical wavelengths but highly absorbent of infra-red radiation at wavelengths around 10 micrometres. This would produce a strong greenhouse effect. Energy from sunlight would go in, be absorbed eventually, perhaps at a depth of 10 to 20 metres, be converted to heat and consequently be unable to escape. The stored energy would raise the temperature of the sub-surface zone. As light would not be reflected significantly at the surface, the surface would appear black.

Hoyle also believes the 'dirty snowball' model of the cometary nucleus to be wrong. He points out that the comet Schwassmann-Wachmann I has a nearly circular orbit a little larger than Jupiter's orbit. Although it never closely approaches the Sun, it displays a violent outburst about once every 15 years. Hoyle believes that there is no possibility that these eruptions could be due to thermal evaporation, but has proposed a biological explanation: that the phenomena of ejection of gas and dust from comets might be the result of gas accumulation due to the metabolic activities of bacteria, leading to high pressures and explosive outbursts.

We mentioned earlier that these two scientists have put forward controversial views on the presence of organisms in space. In their book *Lifecloud* (1978), they proposed that life originated in space, between the stars, that comets are essentially interstellar and might be the site of biopœsis. Organisms 'born' in comets might therefore have seeded other bodies, including the Earth. This was a specific version of the *panspermia* theories discussed earlier. In their book *Diseases from Space* (1979) they proposed that bacteria and viruses reaching the Earth from space might even now be the cause of certain epidemics, a view which has not met with much support from other scientists. In more recent publications, including *Evolution from Space* (1981), *Living Comets* (1985) and *The Intelligent*

Universe (1983) (the last by Hoyle alone), the authors have waded more deeply into realms of fascinating, if not always strongly persuasive, speculation. Hoyle is well known for his fundamental contributions to theoretical astrophysics, and his views deserve responsible consideration. Like other hypotheses, they must ultimately stand or fall on the basis of appropriate tests, which have not yet been done, but which would be possible with present space technology.

Comments on the Hoyle-Wickramasinghe proposals

In the 7 August 1986 issue of *Nature*, Hoyle and Wickramasinghe published an article entitled 'The Case for Life as a Cosmic Phenomenon', in which they defended their interpretation of infrared spectra of bacteria and other materials in relation to observations on the properties of interstellar grains. They pointed out that they started in the 1950s and 1960s with attempts to fit the astronomical observational data to those obtained with inorganic materials in the laboratory, and considered the possibility that mineral and silicate particles in space might be responsible for the observed values. Further work failed to support this suggestion in detail, but the possibility that results with organic grains might provide a better fit was examined, and it then seemed that findings with polysaccharides gave the best correspondence to the astronomical data. They thought it unlikely, however, that large quantities of polysaccharides would have been produced abiologically in space. The authors then realized that dried bacteria might provide infra-red absorption data closely resembling those of interstellar grains. What they considered to be very good agreement was then found between laboratory measurements with bacteria and the average absorption properties of grains along the whole path-length from the galactic centre to the Earth. The measurements were made at 60 wavelengths between 3 and 4 micrometres.

More recently, Hoyle (1986) has discussed the infra-red emission and transmittance characteristics of bacteria and dust expelled by Halley's Comet, and has again concluded that in view of the carbonaceous nature of the bulk of cometary dust, the sizes and densities of the particles and good agreement between observations and laboratory data, 'no person unchoked by prejudice would hesitate to consider very seriously the hypothesis that the bulk of the cometary particles really are bacteria'. Hence his suggestion that cometary outbursts could have a biological explanation.

In response to the criticism that bacteria could not replicate in interstellar space, Hoyle stated that he never claimed that they could. Rather, he envisages bacteria replicating 'as a by-product of

121

star formation in environments where replication can occur'. Later, bacteria would be expelled into space, where some would survive, and Hoyle quotes data relating to the known hardiness of some bacteria in support of his thesis that such organisms could be 'space-hardy'. He believes that some of the properties of bacteria seem more easily explicable as selected for survival in space than for survival on the Earth.

But where does replication occur before these microscopic space voyagers are launched on their fantastic journeys? According to Hoyle, in the interior of comet-like bodies, made warm by radioactive heating, and by the greenhouse-effect of the surface layers. Comets are considered to be 'laboratories' for the development of life and might exist in large numbers around dwarf stars. He calculated that the total mass in the present Oort cloud around our Solar System is about 10^{29} grams, and that the total mass of such bodies for our Galaxy could be 10^{40} grams. These bodies *in toto* therefore could be a vast laboratory for the generation of life, providing a 'more favourable venue for the development of life than the sterile surface of a small planet like the Earth, which even today has a biosphere with a mass of not more than 10^{18}–10^{19} grams'.

The Earth is 'perpetually embedded in a halo of cometary material, of which some 1000 tonnes enter the terrestrial atmosphere each year'. Small objects such as bacteria could make soft-landings and would not be destroyed by flash heating. If comets can be a source of disease-producing organisms, the old idea of the evil nature of comets might have some basis in fact!

Hoyle has been much annoyed by criticism directed at his views by a number of scientists, biologists and astronomers. Some of the criticisms seem to be well-founded, and are referred to and referenced in a letter to *Nature* (6 November 1986), written by R. E. Davies, who points out the possibility that the ultra-violet absorption data for interstellar grains might be accounted for if hollow, spherical carbon molecules, C^{60} (termed *Buckminsterfuller-ene* molecules) were present, a suggestion made by Hoyle and Wickramasinghe themselves, but that the ultra-violet and infra-red spectra of these molecules have not yet been measured in the laboratory. If the interstellar grains are bacteria, the ultra-violet absorption should be much greater than that in the infra-red.

Our own view is that it is not at present possible to discount the suggestion that cometary bodies might be sites in which complex organic molecules and possibly even simple organisms of some kind could be produced. No doubt this will in time be investigated by direct methods. Unfortunately, it is virtually impossible to argue with certainty about the significance of infra-red absorption measurements made on interstellar grains. They might indeed give useful clues to what is present, and perhaps more convincingly an indication of what is *not* there, but with so many unknowns in the situation, firm claims should await the outcome of future space research. Hoyle is always stimulating, and should certainly not be criticized for daring to try to make a contact between astronomy and biology. It seems to us that, on the basis of current information and results, the 'bacteria in space' hypothesis is difficult to defend, but at the same time it has not been shown with certainty to be wrong, and it is not a totally unwarranted speculation. We know that many organic molecules are present in space, and may well be part of the make-up of comets in which greater degrees of complexity could be produced. Against the hypothesis, it has been pointed out that short-period comets may retain their volatile components for only a few thousand years, too short a time to permit biopœsis. This would reduce the number of potentially suitable comets (see Baugher, 1985 and Ponnamperuma, 1981, for more discussion of comets and life).

Meteoroids

We turn next to junior members of the Solar System: the meteoroids. First there are the 'shooting-star' meteors, which are cometary débris, and which can be seen only when they dash into the upper atmosphere and become incandescent. They are small – of sand-grain size – and friable; they do not survive to complete their journey to ground level except in the form of very fine dust.

Meteorites, however, are quite different. They are not associated with meteors, or with comets, and seem to have come from the asteroid belt; indeed, there is probably no difference between a large meteorite and a small asteroid.

The main asteroids are confined to a zone between the orbits of Mars and Jupiter. Only one, *Ceres*, is as much as a thousand kilometres in diameter, and none can ever have retained an atmosphere, and it is unlikely that they could have produced living organisms. There are, however, some small asteroids which may intersect the orbit of the Earth. These so-called *Amor*, *Apollo* and *Aten* objects have been classed as dead comets, which have lost all their volatiles, though this remains a theory, and cannot be proved or discounted until the first successful space mission to one of them.

Many meteorites have fallen to Earth in recorded times, and some of them have produced craters, of which the best-known example is in Arizona. The meteorites are of three main types: *irons* (siderites), *stones* (aerolites) and *stony-irons* (siderolites). Another classification distinguishes *undifferentiated meteorites* (chondrites) and *differentiated meteorites*. In chondrites, the most abundant elements are present in the same proportions as they are in the solar atmosphere. In most cases the ages of meteorites are comparable with that of the Earth. Suggestions that some of them may have come from the Moon, or even Mars, must be regarded as highly speculative, but cannot be discounted.

Meteorites and life

For many years, there have been suggestions made that some meteorites might contain evidence of life elsewhere in the Universe. Failing that, they might provide information on what chemicals similar to those we find in living organisms can be formed abiogenically. Meteorites of particular interest are the *carbonaceous chondrites*, whose composition shows that they probably originated in a part of the Solar System where water or ice was present. Organic molecules are also present in these meteorites. Some may be fragments of comets, perhaps the outcome of a collision between a comet and an asteroid.

On 28 September 1969 a carbonaceous chondrite broke into fragments over Murchison, Australia, and some of the fragments were collected and analysed. Among the organic chemicals found were amino-acids and the bases adenine, cytosine, guanine, thymine and uracil, which are components of nucleic acid molecules. Some of the analyses were performed by Dr Cyril Ponnamperuma, an experienced worker in the field of abiogenic syntheses of organic compounds (see Langone, 1983). It is always very difficult to be sure that contamination of meteorite fragments by terrestrial materials has not occurred, but great precautions were taken to exclude the possibility of contaminants being present in the material analysed. The nature of the molecules present (mixtures of different optical isomers) strongly suggested that they were not of biological origin. In fact, the absence of a predominance of one type of isomer does not in itself completely rule out a biological origin, as over many thousands or millions of years, one optical isomer may slowly change to give a mixture of the D and L forms (the process of racemization).

Work of this kind points to the presence of the 'building blocks' of life in space. The production of these fairly simple organic molecules is not the problem, as Ponnamperuma and many others have shown by extensive work over the years. The important question for biopœsis is how the requisite organization occurred, as we have emphasized earlier.

The presence of *organized structures* in meteorites has also been reported from time to time. Most of these have been shown to be almost certainly terrestrial contaminants. Some of the structures detected in fragments of the Orgueil meteorite, which fell in 1864

124

(see Claus and Nagy, 1961; Nagy and Claus, 1963) were later shown to be probably ragweed pollen grains (see Fitch and Anders, 1963, for a discussion of the nature of 'organized elements' in carbonaceous chondrites). More recently, structures superficially resembling a terrestrial organism, *Pedomicrobium*, in a fragment of the Murchison meteorite, have been reported by Pflug, who also found structures resembling viruses (see Hoyle, 1983). The interpretation of these findings poses enormous problems. It is well known that various inorganic materials will form structures having a superficial resemblance to organisms, and indeed this has been considered by some workers to be a factor pointing to possible prebiological organizational tendencies. A fascinating book dealing with this subject was written at the beginning of the century by Stephane Leduc of the École de Médecine at Nantes, France. An English translation, *The Mechanism of Life*, was published by Heinemann in 1911. (See also Bastian, 1905.) We can only say that work on the presence of organized structures in meteorites is important, but the possibility – even probability – that at least some of the structures detected are non-biological even when they resemble the shapes of organisms, or are terrestrial contaminants, should be firmly considered before sweeping claims are made. Hoyle (1983) has rightly pointed out that the structures found in meteorites, if of biological origin, would be fossils of organisms, rather than living, as the meteorites, which may have been produced as the result of a collision between an asteroid and a comet, would usually have travelled millions or hundreds of millions of times round the Sun before encountering the Earth – a small target. Consequently, any organisms would have been 'alternately roasted and frozen'.

Hoyle (1983) has strongly criticized some of those who have questioned a biological interpretation of these findings. We should be happy to know that there are remains of true organisms in meteorites (as there may be) but the finding would be of such profound significance that it is reasonable to demand the highest possible standards of supporting evidence. Calm, informed and unhasty consideration of the evidence is desirable, and we agree that there is a case for further investigation. One can sympathize with careful workers who feel that their labours are not adequately appreciated by colleagues, but they are investigating a particularly difficult and treacherous territory, and must be prepared for searching criticism. Future, more rigorous work and the application of new techniques of analysis should help to solve some of the problems encountered in investigations of this kind. A review of organic matter in the Orgueil meteorite was given by Baker (1970–1), who concluded that it was probably not of biological origin.

IX PLANETS OF OTHER STARS

We have given considerable attention to the possibility of finding life on other planets or on satellites in the Solar System, and have concluded that, apart from on Earth, it is most improbable that life exists. When we consider life in the Universe, our Solar System is only part of the question – and a very minor part at that. The Sun is a normal star, and other stars almost certainly have planetary systems of their own.

Formation of the Solar System

Before we can decide how many other stars are likely to be attended by planets, we must consider how our Solar System was formed. In 1755, the philosopher, Immanuel Kant, advanced a 'gas-cloud' theory, and the famous Nebular Hypothesis proposed by the mathematician Laplace in 1796, also involved a cloud of material which was assumed to surround the youthful Sun and from which the planets condensed. Mathematical difficulties led to the rejection of Laplace's theory during the nineteenth century, and it was replaced by a crop of 'catastrophic' hypotheses involving an encounter between the Sun and a passing star. The most famous exponent of this idea was Sir James Jeans, who adopted the theory in our own century and popularized it in books and broadcasts (see Jeans, 1930).

If the planets were pulled off the Sun by the action of a wandering star, then solar systems would be very rare indeed, because close stellar encounters seldom occur. Indeed, our Earth might be the only planet of its kind in the entire Galaxy. However, Jeans' and other catastrophic theories have also been found wanting, and today we are confident that the planets grew up by accretion from a *solar nebula*. Near the freshly-luminous Sun, the planets lost most of their

126

volatile elements; further out, substances such as hydrogen could be retained, producing the giant planets.

We know the ages of the Earth and Moon fairly precisely (around 4600 million years) and the same is no doubt true of other planets, so that the minimum possible age of the Sun must be put at 5000 million years. At the moment, the Sun is in a stable condition and has been so for a period long enough to permit the evolution of advanced life forms such as ourselves. Neither is it expected to change markedly for several thousands of millions of years in the future.

Other stars, however, behave differently. Everything depends upon their initial mass, and in particular the time-scales are by no means the same.

Stellar classification and evolution

The stars are classified by their spectral characteristics. Those of types W, O, B and A are hot and white or bluish-white; types F and G are yellowish; K orange and M, R, N and S orange-red. The series indicates decreasing surface temperature, from at least 80,000°C for type W down to only about 2500°C for N and S. For our present purpose, we need consider only the middle part of the series, from B to M. Stars of the remaining types are comparatively rare.

In 1911 the Danish astronomer, Ejnar Hertzsprung, plotted the stars on a diagram according to their luminosities, in terms of the Sun, and their spectral types (or surface temperatures, which comes to the same thing). Similar work was carried out in the USA by Henry Norris Russell, and the resulting Hertzsprung-Russell or HR diagrams are of fundamental importance to astrophysics. It is found that most of the stars lie in a well-defined band extending from the upper left to the lower right; this is termed the Main Sequence. There is also a 'giant' branch, to the upper right. It is evident that the cooler stars (K and M) are divided sharply into highly luminous giants and feeble dwarfs; the distinction is less clear for the yellow stars, and vanishes for those with the highest surface temperatures. The 'white dwarfs', to the lower left of the diagram, come into an entirely different category.

Originally, it was believed that the HR diagram indicated an evolutionary sequence, with a star beginning its career as a large, diffuse red giant, joining the Main Sequence after contracting by gravitation, and then moving down the Main Sequence from left to right, ending up as a cool, dim red dwarf. Later work showed that this was incorrect, and the true sequence of events is rather different.

A star with less than 0.1 solar mass (solar mass = mass of the

Sun) will condense out of the material in a nebula, but will never become hot enough at its core for nuclear reactions to begin (about 10 million °C is the 'starting point'). Therefore the star will simply shine feebly because of gravitational contraction until it becomes a cold, dead globe. Stars of this kind are called, rather misleadingly, brown dwarfs. In 1984 D. McCarthy and F. Low, in America, reported the detection of such a body, associated with the dim star, Van Biesbroeck 8, which is 21 light-years from us. This announcement caused considerable interest, but unfortunately it has not been confirmed, and we have still not identified with certainty any brown dwarfs, though in theory they should be plentiful.

Stars such as the Sun – with a mass between 0.1 and 1.4 times the solar value – will begin by condensing gravitationally out of nebular material. The star will contract, decreasing in luminosity, until it joins the Main Sequence and nuclear reactions begin at the core. As we have seen, the main process is the conversion of hydrogen to helium. Condensation to the Main Sequence takes several million years, but once there the star remains stable for something like 10,000 million years, so that our Sun is no more than half-way through its Main Sequence career. When its supply of hydrogen 'fuel' is used up, the star has to change its structure. The core shrinks, with a tremendous increase in temperature, and the outer layers expand so that the star turns into a red giant. Subsequently, when all the nuclear reserves are exhausted, the star collapses into a very small, super-dense white dwarf, in which all the atoms are broken up and crushed together, and occupy much less space than previously. The best-known example of a white dwarf is the faint companion of Sirius, with a mass similar to that of our Sun but in size smaller than the planet Uranus – so that its density is at least 60,000 times that of water. A white dwarf has been described as a bankrupt star, and will end its career as an inert globe.

With stars of mass greater than 1.4 that of the Sun, the pace of evolution is accelerated. After the end of the Main Sequence period, different reactions take place in the core, and the final result is an 'implosion' (the opposite of an 'explosion'), so that the star literally blows itself to pieces in what is termed a *supernova* outburst. The end product is a cloud of expanding gas, which gradually dissipates, while the remnant of the old star is left; it is composed of neutrons, and is far denser even than a white dwarf, and has a diameter of a few kilometres. It spins rapidly, producing the rapidly-varying radio emissions which have led to neutron stars being named 'pulsars'. Gradually the emissions become less frequent, and finally cease.

With a still more massive star, there is no supernova outburst, and after its abrupt collapse the stellar remnant is left with such a strong gravitational field that not even light can escape from it. It is surrounded by a region which is to all intents and purposes cut off from the rest of the Universe; it is called a *black hole*.

The time scale is a most important factor in our search for other possibly inhabited planetary systems. With a solar-type star, we can assume, by analogy with our own Solar System, that there will be plenty of time for life to evolve as it has on the Earth, should a suitable planet exist; but with a more massive star, this is not the case. For example, consider the very hot star Theta Carinæ, in the far south of the sky. Its age is estimated at no more than 10 million years; and even if it has associated planets, the development of life in the Theta Carinæ system is most unlikely. This will also apply to a typical A-type star such as Altair (Alpha Aquilæ), where the estimated age is less than a thousand million years.

It seems, then, that we must concentrate on stars of spectral types between types F and K, with emphasis on those of type G (the Sun is a typical G2 star). With these, we have time-scales of the right order, together with masses which are comparable with those of the Sun. We are, of course, dealing here with stars on the Main Sequence, and excluding the F-type supergiants such as Canopus and the K-type giants such as Arcturus.

Detection of extra-solar planets

Unfortunately, no telescope yet built is capable of showing even a large planet associated with a relatively nearby star. The Hubble Space Telescope, due to be launched into space in the foreseeable future, may be able to do so, but even so it will be limited to planets of Jovian size, many of which, if they exist, may not be promising sites for habitation by intelligent beings. For the moment, we are limited to two possible means of investigation: astrometry and infra-red measurements.

If a relatively lightweight star is attended by a massive planet, there may be detectable effects upon the star's proper motion. The principle is clear enough; thus the 'wobbling' of Sirius against its background was studied by F. W. Bessel in the 1840s, and given as proof that the star must be perturbed by an invisible companion. In 1862 the companion was discovered optically by the American telescope-maker, Clark, and is of course the famous white dwarf, Sirius B. But Sirius B is comparable in mass, though not in luminosity, with Sirius itself (the ratio is less than 1:3), and so the perturbations are marked. With a planet, which is much less massive than a star, the situation is far more difficult.

Nonetheless, investigations carried out chiefly at the Sproule Observatory by P. van de Kamp (see van de Kamp, 1987) have indicated that Barnard's Star, a faint red dwarf at a distance from us of 6 light-years (and thus the nearest star beyond the Sun, apart from the triple Alpha Centauri system), shows perturbations which could be due to a planet or planets. If these observations are correct – and recently doubts have been cast on them – then the Barnard's Star attendants are comparable in mass with Jupiter. A much less massive world, such as the Earth, could produce no measurable perturbations. Barnard's Star is not the only case cited by van de Kamp, but at the moment the astrometric method cannot be regarded as at all reliable.

Information of a different type was obtained from IRAS, the Infra-Red Astronomical Satellite, which was launched in January 1983 and remained operating from orbit for nearly a year. More infra-red radiations from space are blocked out by the Earth's atmosphere (which is one reason why astronomers try to make the best of things by going to high mountain peaks; the world's largest infra-red telescope is sited at the crest of Mauna Kea, in Hawaii, at nearly 14,000 feet) and therefore specially designed satellites are invaluable. IRAS was particularly successful, and led to some discoveries which were as unexpected as they are fascinating.

Data from IRAS were received at the Rutherford-Appleton Laboratory in Oxfordshire. It was from here that two American members of the team, H. Aumann and F. Gillett, were testing the infra-red telescope by calibrating it on certain typical stars when they found that Vega or Alpha Lyræ, the fifth brightest star in the sky, had what Gillet called a 'huge infra-red excess'. The source was about 20 seconds of arc in diameter, and clearly indicated the presence of cool material. Further studies showed that the material took the form of a cloud of particles, many of them larger than those normally found in interstellar space. Moreover, the region extended out to a distance of some 80 astronomical units (an astronomical unit is the mean distance between the centre of the Earth and the centre of the Sun) from Vega, and it was estimated that the total mass of the cloud was much the same as the masses of all the planets in our Solar System. Presumably there were particles much larger than dust grains, because very small pieces of material would have been drawn back into the star – leaving intermediate and large-scale debris in orbit. In fact, there seemed to be a strong possibility that some of these pieces of debris could be of planetary size.

The hunt was on. Some other stars, including Fomalhaut in the constellation of the Southern Fish, showed similar infra-red excess. Altogether some sixty examples were found before IRAS came to the end of its active career. Of special interest was Beta Pictoris, in the far south of the sky. The infra-red excess was unusually great, and it seemed worth while to try to observe the material at optical wavelengths. Using the 254-cm. Irénée du Pont reflector at the Las

Campanas observatory in Chile, Bradford Smith and Richard Terrile succeeded in obtaining pictures. The material extends on either side of the star out to a distance of some 400 astronomical units, with a maximum thickness of 100 astronomical units; the appearance appears to be due to a disk of material, and we are observing it almost edgewise-on. The star itself is dimmed by only half a magnitude, indicating that the debris does not extend right down to the surface – and it is in this disk of debris that planets could be formed. In fact, Terrile went so far as to claim that 'it would be very hard *not* to form planets from material like this'.

Obviously we cannot yet form any definite conclusions, but the evidence is mounting up. If many of the stars in our immediate neighbourhood have such systems, then we cannot reasonably doubt that they are common throughout our Galaxy and others. This is not to suggest that every planetary system must contain inhabited worlds, as will be clear from the discussion of the problems of life's origin in other chapters. It is worth noting that Vega, Fomalhaut and Beta Pictoris are all hotter, more energetic and younger than our Sun. Data are as follows:

	Fomalhaut	Vega	Beta Pictoris
Spectral type	A3	A0	A5
Luminosity (Sun = 1)	13	52	40
Distance (light-years)	22	26	78
Estimated age (millions of years)	400	230	300

Although prokaryotes may have appeared on Earth quite soon after its formation, evolution to produce intelligent life took thousands of millions of years, and it seems unlikely that more than a beginning in biological activity would have occurred on planets in the three systems listed above, should suitable planetary environments for the emergence of life exist. Yet the results are significant. Though we cannot prove that planets accompany any of the sixty known infra-red excess stars, we can be fairly confident that potentially planet-forming material does.

The Sun is a very 'normal' star, and indeed G-type dwarfs are among the commonest in the Galaxy. Other types of stars may be less suitable. We can exclude the variable stars, and probably most binary systems also; stars which have left the Main Sequence and have reached the giant branch of the HR diagram will probably have destroyed any attendant planets which might have existed.

Contacting other civilizations

The only means of which we are aware by which communications with other worlds might be achieved is by use of signals coded in electromagnetic radiation. Radio is available to us now, and in future laser beams may offer an alternative. If other civilizations were to attempt to contact us, it is conceivable that they might use these methods, but the possibility exists that techniques of which we are at present totally ignorant might be employed.

In identifying possible life-bearing systems with which we might communicate, it seems reasonable in the first instance to restrict ourselves to certain stars of types F to K that are not excessively distant. As electromagnetic signals in space travel at the speed of light, if we send a message to a star 30 light-years distant, we cannot receive a reply in less than 60 years. For convenience, it seems appropriate to list first those seemingly suitable and probably planet-attended stars within a radius of 30 light-years. On the basis of current information, fewer than 20 potentially suitable stars can be identified (see Table 2).

Table 2

Star	Spectrum	Distance (light-years)	Luminosity (Sun = 1)
Epsilon Eridani	K2	10.7	0.3
Epsilon Indi	K5	11.2	0.2
Tau Ceti	G8	11.9	0.4
Sigma Draconis	K0	18.5	0.3
Delta Pavonis	G5	18.6	1.0
e Eridani	G5	20.2	0.4
Beta Hydri	G1	20.5	7.0
Xi Bootis	G2	22.0	0.8
Zeta Tucanæ	G0	23.0	0.8
Chi Draconis	F7	25.0	2.0
Pi^3 Orionis	F6	25.0	3.0
Gamma Leporis	F6	26.0	2.0
Mu Herculis	G5	26.0	2.0
Beta Comæ Berenices	G0	27.0	1.2
61 Virginis	G6	27.0	0.7
Beta Canum Venaticorum	G0	30.0	1.3

Of the three nearest, Epsilon Indi is possibly too underluminous, and the best candidates are Epsilon Eridani and Tau Ceti. Both are much less luminous than the Sun, but the astrometric method has indicated that planets may be present, and in 1960 they were the subject of a novel experiment.

If we hope to attract the attention of an extraterrestrial civilization by means of radio, we must select a suitable wavelength, and the ingenious choice of 21.1 cm. was made. This is the wavelength of radiations emitted by clouds of cold hydrogen spread through the Galaxy. It was predicted by H. van de Hulst in 1944, and confirmed experimentally by Ewen and Purcell after the end of the war. If other races using radio exist, and are as scientifically advanced as ourselves, they should be aware of emissions from interstellar hydrogen, and might be studying them with their own radio-telescopes. They might therefore notice patterned signals transmitted at this wavelength.

Two suggestions were put forward. It might be thought worth while to make regular transmissions at 21.1 cm. in the hope that some far-away operator would receive them and reply; alternatively, we might 'listen out' and see whether we could detect any signals regularly patterned enough to be regarded as probably artificial. Work along these lines was started in 1960 by radio-astronomers at Green Bank, West Virginia, using the large radio-telescope there. It was the 'longest of long shots'; not surprisingly the results were negative, and Project Ozma, as it was officially called (or more colourfully by the astronomers involved, 'Project Little Green Men'), was discontinued after a few months, but it was at least a first attempt.

Since then, other experiments of the same kind have been put in hand, and the Search for Extraterrestrial Intelligence (SETI) has been introduced. Though the chances of success are impossible to estimate, the idea is not irrational, and *if* there are accessible civilizations, communication is in principle possible. The belief that the principles of mathematics would have had to be discovered and used by any technologically advanced civilization has led to suggestions on codes that might have a universal quality, but the subject of the language of communication is probably more difficult than it may appear at first sight.

For the present, there is no realistic possibility of interstellar manned expeditions, although some scientists believe that it is in principle possible. We have no way of knowing what might be the attitude toward us of any visitors who might come from other systems. Even if they are peaceful to each other, they might find humans primitive or displeasing. Wide separation of planetary systems might have its good points! Interesting though these matters are, further discussion would take us too far from our main theme.

X ALIEN LIFE

So far, we have confined our discussion mainly to life of the sort familiar to us. There are, however, broader concepts which must be dealt with in some detail.

Let us suppose that somewhere in our Galaxy (or, for that matter, in another galaxy) there exists a world on which conditions after its formation were essentially similar to those on Earth, and that it has a similar gravity and comparable radiation from its parent star. If life has developed on that world, it will have undergone evolution. As chance events can have a profound influence on the future development of systems far from equilibrium, we might be incorrect if we supposed that humans or human-like creatures would be produced, even given a biochemistry similar to our terrestrial kind. Intelligent beings might indeed develop, and selective advantages of some features might lead to resemblances in form between some 'other-world' and terrestrial organisms. However strange the 'other-worlders' might appear to us, they would be built of fundamentally similar materials, and even if they looked like some of science fiction's 'bug-eyed monsters' (BEM's), they would not be truly alien, except in appearance and location of their habitat.

Fantastic ideas

When we depart from established scientific principles we enter the realm of fantasy (not necessarily a bad thing to do), and meet creatures such as the Martians described by H. G. Wells in his classic novel *The War of the Worlds*. His own account of them reads as follows:

> They were, I now saw, the most unearthly creatures it is possible to conceive. They were huge round bodies – or, rather, heads – about four feet in diameter, each body having in front of it a face.

This face had no nostrils – indeed, the Martians do not seem to have had any sense of smell – but it had a pair of very large dark-coloured eyes, and just beneath this a kind of fleshy beak. . . . In a group round the mouth were sixteen slender, almost whip-like tentacles, arranged in two bunches of eight each. . . .

The Martians were very dependent on the machines they used for locomotion and for handling objects.

Wells' story is good reading and entertaining, but it is important to note that his fantasy was contained by his knowledge of the science of the time. The Martians were built on the same principles and of the same materials as we are, but had evolved differently, perhaps having been derived by evolutionary change from beings like ourselves. They were mostly brain, and did not eat, but injected themselves with blood from other living beings. This was conceived by Wells as a physiological economy, freeing the Martians from the necessity of carrying out the processes of digestion. They were also asexual, and so 'without any of the tumultuous emotions that arise from that difference among men'. New Martians were budded off from the parent. The tentacles were used to work machines that replaced lost anatomical and physiological features, so that the Martians were 'practically mere brains, wearing different bodies according to their need', and their machines were 'actuated by a sort of sham musculature of disks in an elastic sheath; these disks become polarized and drawn closely and powerfully together when traversed by a current of electricity'. This foreshadows present-day suggestions that technology might be able to make use of systems in some ways based on biological mechanisms. The death of the Martians, and of their 'red weed', a Martian plant that was brought with them, was apparently the result of bacterial infection, the Martians having no immunity to Earth germs, and seemingly no microbial diseases of their own. The story is all the better for retaining a foothold in established terrestrial science. Few authors of science fiction write with such skill, and some of the recent BEM's are disgusting without being entertaining or thought-provoking.

In our opinion, there is no good evidence that the Earth has been visited by beings from other worlds (as some writers seriously maintain) whether in flying saucers or other craft, or by teleportation. We cannot say for certain that this is so, but the onus is on those who maintain that visitations have occurred to produce acceptable evidence – even one good, unfaked photograph. We are also ready to admit that visitations might be difficult to prove, might catch observers off-guard, and that the visitors might take steps to remain elusive. Unfortunately, eye-witness accounts of what appear to the observer to be unusual events are too often unreliable, and lack of good evidence is not cause for belief. The best we can do at present is keep an open mind.

Notions of alien life

In general, it can be said that there are several somewhat different aspects of the problem of the existence of alien life. If the Universe should contain two or more forms, based on different physical structure or radically different 'biochemistries', then what was to be considered 'alien' would depend on which system was making the judgement. To thinking, silicon-based organisms *we* would appear to be alien. It is even possible that our form of life might be the exception rather than the rule.

In a narrow sense, we can speculate on the possibility of the existence of beings not unlike Earth organisms on other planets. The most closely similar would be organisms based on the same basic chemistry but making use of key molecules, such as amino-acids, that are mirror images of the form found in terrestrial organisms. So far as we know, there is no fundamental reason why organisms elsewhere should not be of this kind, but it is important to remember that the reason for the selection of the particular molecular optical isomers in terrestrial biology is far from clear. It might be the result of chance factors operating at the time of formation of some initial critical molecular complex which was then replicated rapidly enough to take over from competing systems. There is, however, a possibility that for reasons unknown certain molecular forms are essential for life. Unsuspected asymmetries have been discovered in physics, and these may be reflected at other levels of organization. It seems not unreasonable to suspect that on at least some other planets around distant stars, carbon-based organisms are likely to exist. Because of the large number of parameters that must be held within narrow limits to permit the emergence and existence of life, life-bearing planets in this sense may be rare and separated by huge distances.

Within the same general framework, we can speculate on the possible existence of organisms in which carbon remains a critical component but in which water has been replaced by some other solvent. The most usual suggestion is that liquid ammonia might replace water, but as was explained earlier, water seems to have some unique properties and may well be essential for any form of carbon-based life. The hope of finding, in the atmospheres of the outer planets or on their satellites, carbon-based organisms living at low temperatures and utilizing liquid ammonia (at a temperature around −53°C) in place of water has been expressed by several authors, but the likelihood of this seems to be extremely low. It has been suggested, too, that at low temperatures nitrogen atoms might form chains and act in ways analogous to the carbon in terrestrial organisms. Nitrogen chains would be unstable in our environment, but might be able to persist at low temperatures, perhaps forming mixed chains with carbon atoms. Unfortunately, it is not enough for an alternative biology that this or that element might behave in an appropriate fashion. Many different things have to be 'right' simultaneously, and we may suspect that the possibilities for self-organization and selection of successful systems are not unlimited. The most we can say at present is that, when the outer planets and their satellites are studied by sophisticated probes in the future, a search for alternative life forms would be justifiable.

In science fiction, and at times in serious scientific speculative writings, *silicon* has been suggested as a possible substitute for carbon. A superficial comparison of the carbon and silicon atoms reveals certain promising similarities (each has 4 electrons in the outermost shell), and both are able to form chains (Fig. 22). The chains of silicon atoms are, however, shorter and less stable than those of carbon. Carbon and hydrogen form methane, CH_4, silicon and hydrogen form a silane, SiH_4. There are large numbers of carbon compounds which have counterparts in silicon chemistry – for example C_2H_6, Si_2H_6; CH_3OH, SiH_3OH – and there are some mixed compounds such as CH_3SiH_2OH.

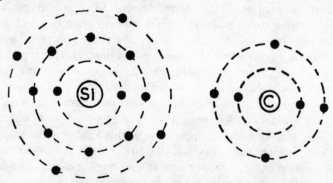

Fig. 22 Electronic structure of silicon (Si) and carbon (C) atoms

In spite of these similarities, there are important differences between silicon and carbon which may rule out the existence of silicon-based organisms. The silicon atom has a greater radius than the carbon atom, the 4 outermost electrons being in the third shell, whereas in carbon they are in the second. Consequently, the outermost electrons are less strongly attracted by the nucleus than in carbon. Molecules of carbon dioxide, CO_2 do not link together, but when silicon is combined with oxygen, the weaker attraction of the nucleus for the outermost electrons permits an oxygen atom to be shared between two silicon atoms, leading to the formation of a crystalline solid in which the atoms are firmly bound. Silicon dioxide is therefore quite different from carbon dioxide, for it is a solid with a very high melting point and low solubility in water, occurring naturally in the form of sand, quartz and fused silica. In carbon-based organisms, carbon dioxide performs essential functions, and clearly silicon dioxide, which is gaseous only above 2500°C, could not play a similar rôle in water-dependent organisms.

The capacity of the outermost shell of the silicon atom to accommodate up to 18 electons, whilst that of carbon can accommodate at most 8, introduces further differences. When carbon is sharing electrons with each of four other atoms, the resulting 'octet' of electrons around the carbon nucleus is not able to 'expand' to receive further electrons. The related silicon compounds are more open to attack, for instance by water, so that the -Si-Si-Si- sequence can be changed to -Si-O-Si-. The strong carbon-carbon bonds which can be formed make possible an enormous variety of stable organic compounds.

The ability of silicon to form complex molecules in association with oxygen, and some other elements (e.g. aluminium and phosphorus) has been hailed by some authors as a possible basis for alternative life forms at high temperatures, and Feinberg and Shapiro have proposed the term *lavobes* for such hypothetical microbial entities whose home would be molten rock, and postulated the possibility of evolution of the systems. The same authors have further suggested that the Earth might harbour this type of life, and have proposed the term *magmobes* for organisms of the Earth's magma, beneath the crust. If organisms of this kind exist, they would be expected to be present on volcanic planets or satellites, and Jupiter's satellite, Io, would be a possible site in our Solar System. The idea of mineral organisms of this kind may be regarded as supplementary to the proposals of Cairns-Smith discussed earlier.

Under great pressures, water and silicon might co-exist in a liquid state, but exactly what would happen under these circumstances is uncertain. Wilder speculations about silicon organisms living at furnace temperatures and thinking at lightning speed have been tempered by consideration of the numerous conditions that would have to be met to permit an alternative biology. There is, however, a distinct possibility that organization at the mineral level might generate systems that have some lifelike attributes, and there might be a sense in which these are our ancestors, setting the stage for later organization of carbon-based life. It is also possible that systems of this kind are more widely distributed in the Universe than are carbon-based organisms, and that only exceptionally do they have the opportunity to take part in the initiation of the sort of life with which we are familiar.

Sulphur and oxygen

In the same way that the silicon atom resembles the carbon atom, there is a resemblance between oxygen and sulphur atoms. In the oxygen atom, 2 electrons are in the first shell and 6 in the second. In the sulphur atom, there are 2 electrons in the first shell, 8 in the second and 6 in the third. Each of these elements has 6 electrons in its outermost shell, and this determines certain chemical similarities; but as with carbon and silicon, there are important differences, partly arising from the ability of the outermost shell of the sulphur atom to accommodate more than 8 electrons. It seems unlikely that sulphur could play a role closely analogous to oxygen, as has sometimes been suggested. Sulphur is an important element in carbon-based organisms, and numerous biochemical reactions occur in which it is involved. It is possible that under different conditions it might be able to take part in some novel high-efficiency metabolic system.

To the fringes of speculation – and beyond?

The generalization of the definition of life to cover systems quite unlike any with which we are familiar (see p. 56) has led to a number of specific suggestions on the feasibility of development of physical systems with life-like characteristics ('physical life'). We have already mentioned the possibility of constructing intelligent, self-reproducing electronic devices, and work in the field of artificial intelligence is progressing rapidly. The man-made devices would be a product manufactured by carbon-based organisms, and we might draw a comparison here with the possible ordering of carbon compounds by mineral organisms. In the same way that carbon-based organisms and silicon-chip based reproducing systems might co-exist, so might mineral life and carbon-based life. Now, however, it is time to move beyond the range of lavobes and magmobes, and to consider physical systems that might be generated in environments usually considered totally hostile to any form of life, as in the interior of stars or in space. This is usually thought of as a hunting ground for the writers of science fiction, but events of the last few decades have demonstrated that truth has in fact often turned out to be stranger than fiction. Whether this will be so in the field of life in the general sense, remains to be seen.

The bio-enthusiasts Feinberg and Shapiro have used the term 'plasmobes' to refer to systems that might occur in certain zones of a star, where there are no molecules but there is a plasma consisting of positively charged ions, electrons and intense magnetic forces. Could ordered structures, able to replicate, be produced under these conditions? They would have to consist of patterned magnetic forces and electric charges. The existence of 'living' entities of this kind had been suggested earlier by A. D. Maude, to whom Feinberg and Shapiro refer. The basic problem appears to be the degree of probability of producing a fairly simple replicating magneto-electric system, a situation which can be likened to the need to produce an Eigen-type hypercycle or something similar to initiate carbon-based life. In principle, the emergence of ordered patterns in some regions of plasma would be understandable in the light of Prigogine's work on the production of order from chaos. Stellar energy, flowing from the deeper layers outwards, would be available to drive the production of dissipative structures if suitable deviations from equilibrium and fluctuations occurred. It is possible, too, that plasmobes might have a wide range of size, from microscopic to astronomical dimensions. Could such life develop 'purposive' aspects and come itself to exert an element of control on events within stars, even as carbon-based life exerts effects on our planet? The implications of this possibility are remarkable. Feinberg and Shapiro also refer to the possibility of what they term 'magnetic atom polymer life' at the surfaces of neutron stars: atoms polymerizing in the intense magnetic fields. They speculate that organisms existing in the enormously strong gravitational fields associated with neutron stars would be thin, although possibly widely spread out. It is apparent that the investigation of the possible existence of organized systems of the type described – plasmobes, 'plasmobeasts' and others – would be technically extremely difficult, and the problem of recognition might be great. These are systems that might function at very high rates, but whether a means of meaningful interaction with them could be devised, we do not know.

Life in stars would be very high temperature life, but speculation has also ranged over the other end of the scale – the possibility of finding life at temperatures not many degrees above absolute zero. Our indefatigable space biologists, Feinberg and Shapiro, have proposed the name '*H-bits*' for organisms made mainly of solid hydrogen and inhabiting very cold planets remote from their sun. The activities of these organisms would depend on molecular rotation of o- and p-hydrogen. An H-bit is pictured as having a liquid or gaseous centre of helium, surrounded by a layer of o-hydrogen which in turn is surrounded by a layer of o-hydrogen containing some magnetic impurity molecules. The H-bit would function by catalysing the production of p-hydrogen from o-hydrogen, and the reverse reaction. Mechanisms for replication are discussed. Solar radiation would be the source of energy to produce order.

The possibility of life in interstellar space has been approached from different angles. At one level, as extensively discussed in the works of Hoyle and Wickramasinghe during the past few years, we are dealing with the existence of materials and organisms essentially like those we know, able to seed planets and being in fact the source of life on Earth. Could other materially different but in a general sense 'living' systems exist in space? Interstellar clouds have provided food for speculative thought, and the existence of a *radiobiosphere* has been postulated. Stellar radiation would be the source of energy, and photons would be absorbed by the clouds. Feinberg and Shapiro have provided us with yet another term, *radiobes*, for the individualized entities constituting radiant life, which would manifest as ordered patterns of radiation emission by excited atoms. Atoms would be necessary for radiant life, as radiation does not in general interact with itself. According to Feinberg and Shapiro, matter could be used by radiation to produce order in itself. An excess of matter would lead to many matter-matter interactions that would prevent the formation of radiobes, but interstellar space with sparse matter and much radiation would be a suitable site for their development.

As Feinberg and Shapiro modestly conclude, 'we cannot be certain that radiant life exists in the Universe. The same is true for the other forms of physical life we have discussed.' We feel, however, that the discussion of these possibilities is stimulating and could in time lead to a more 'organic' picture of the Universe at large, perhaps to some extent in keeping with the intuitions of certain philosophers (e.g. Whitehead) and philosophically inclined physicists. Even if life as we know it is thinly spread in the Universe, we may reasonably suspect that there is a continuity between the 'inorganic' and the 'organic', and this would be true even if a critical event should be necessary to initiate carbon-based life, an event of such low probability that the Earth is the only abode of this type of system. On the other hand, the possible existence of other more or less 'life-like' systems based on different organizations of matter and energy should not be disregarded. Mineral organisms may well exist on the Earth now, but if they do, we have been unaware of them. To us, with our limited knowledge of the constitution of the Universe, consciousness seems extraordinary, if not almost miraculous. Are there different modes of consciousness, associated with different organizations of matter and energy? How, if at all, could communication between one type and another be possible? What would the experiences of one type of system 'mean' to another? Here we are still almost in the realms of science fiction, and can do no more than acknowledge the fascination of many as yet unanswerable questions.

XI OVERVIEW

In the closing chapter of the first edition of this book we stated that
'it seems probable that the general direction of thought on the
problem of biopœsis is more or less correctly established.' This
referred to the production of carbon-based organisms of the type
with which we are familiar. In the light of theoretical advances and
new discoveries it now appears that this statement was true in only
a limited sense. There is still good reason to believe that some of the
various ways in which organic compounds can be formed from
simpler materials are important for the production of organisms,
whether or not the Earth itself was the site of production of living
things. We can see, however, that organisms can be regarded as an
integral part of the Universe, and that a large number of very special
conditions have to be correct for their existence. The whole story of
the origin and development of our kind of Universe with its
particular laws of physics seems therefore to be crucial for the
production of time scales and environments in which organisms can
develop.

Some, but not all, physicists have become increasingly impressed
with the possible creative rôle of observer 'participation' in the
Universe, and a few would argue that the evolution of consciousness is what in fact introduces 'meaning' into the Universe. We can
suspect that life is not just an interesting accident in the Universe,
even if it is to be found only very thinly distributed or on only one
planet. It is part of a gigantic process of which we at present may
have only the most rudimentary understanding. If the Earth is in
fact the only inhabited world, that does not render life unimportant
or insignificant, in spite of our planet's minuteness in relation to the
vast extent of the Universe. It has been suggested that to give a
realistic chance of producing the remarkable phenomenon of carbon-
based life in even one small locality, a precisely constituted Universe

as big as ours would be necessary. From this perspective, the assembly of the parts of living organisms may seem less remarkable than the building of the factory.

Carbon-based life, as we have seen, may not have arisen directly from mixtures of organic chemicals in a primordial soup, but the organization of organic molecules may have required the prior formation of 'mineral organisms' of the type postulated by Cairns-Smith. This is an attractive idea, because it provides a basis for understanding how essential features of organization might have been initiated at the dawn of the era of carbon-based life, and we have seen how the emergence of technologies developed by carbon-based life has led to the production of devices such as silicon chips and the possibility of producing new types of self-reproducing, potentially intelligent 'machines' which might not unreasonably be included in a general category of 'organisms'. In the same way that carbon-based life probably did not arise unaided from a soup, so these 'machines' would not have sprung directly from the rocks. Their production has been initiated by carbon-based organisms. For the task of exploring the Universe, examining its qualities and, if the 'participationists' are correct, giving its process direction and form, rugged electronic devices might have greater flexibility of tolerance of different environmental conditions and be more enduring than carbon-based organisms. Whether they could in fact ever develop consciousness, feelings or emotions is at present an unanswerable question. If they could not, they would still be invaluable as exploratory devices for the enrichment of the stores of knowledge acquired by carbon-based organisms.

It was argued in Chapter X that silicon could probably not replace carbon as the basis of a life form analogous to carbon-based organisms, and that other solvents, such as liquid ammonia, could probably not replace water. Carbon and water have too many special features to permit easy replacement, and changing one or two components would make necessary numerous other modifications. It is hard to imagine environments in which silicon-based organisms or 'ammonia organisms' would have a realistic chance of developing and persisting. Could 'physical life', in the form of Feinberg and Shapiro's 'plasmobes', 'radiobes', etc., exist? It may be that various types of organization of the type discussed are possible, but to what extent they would in fact be comparable to carbon-based life, except in a rather general thermodynamic sense, is hard to know. Whether or not they should be called 'living' is perhaps a matter of definition, but a catalogue of what they could actually do would be informative. If such systems do exist, they must play a role in the overall organization of some aspects of the Universe, even if not in the same way that carbon-based life does.

Table 3 Postulated sequence for development of organisms

1. The Big Bang. Quarks → Nucleons.
   (During the first 10^{-6} sec.)
2. Universe contained protons, neutrons, electrons, neutrinos and photons during the first second of its existence. Intense radiation level prevented formation of atomic nuclei.
3. After one second, temperature fell. Formation of deuterons, then helium-4 nuclei, some helium-3 and lithium-7 nuclei. This was the phase of primordial nucleosynthesis, which lasted a few minutes.
4. Phase of cooling (10^6 years) followed by formation of galaxies and stars. Stars are site of further nucleosynthesis, including nuclei of elements necessary for carbon-based life.
5. Stellar explosions (supernovæ) scatter elements into space.
6. Formation of molecules in interstellar dust, comets, and on planets of stars. Planets may have acquired complex molecules (even perhaps organisms formed on other bodies?) from space.
7. Growth in complexity of molecules and molecular systems. Chemical evolution. Perhaps production of 'mineral organisms' before switch to carbon-based life.
8. Production of replicating systems (pre-carbon based or carbon-based) made possible evolution of neo-Darwinian type. Novelty and increase of information in the developing organisms arose by selection acting on systems produced by 'chance'. Intelligent carbon-based on life on Earth 'culminating' in humans.
9. Humans are in the process of developing new systems that may qualify as 'intelligent'. Self-reproducing computer-controlled systems are a possibility. Will these be in effect another form of 'organism', supplementing or even supplanting carbon-based life if conditions become unfavourable for its existence?

Are there influences directing the development of the Universe, as suspected for example by Hoyle, and making possible events that seem otherwise to be so improbable as to be virtually impossible? It is not possible to answer this question with certainty, but the Universe does seem to be so constituted that what we consider to be 'order' is produced, and we have seen how the very fact that the Universe is expanding is a necessary aspect of this capability. The notion of 'inventiveness', and even perhaps of a sort of 'intelligence', may not be wholly inapplicable to the Universe itself, and the recent demonstration of large-scale coordination of behaviour of constituents of the Universe by means as yet unknown points to the existence of constraints that may have important consequences for the organization of matter.

How did the Universe come into existence? Some would say as a result of a fluctuation in the quantum vacuum, which itself may be conceived as a seething field of activity with the continual production and annihilation of 'virtual particles'. Indeed, it may be that the total energy of the Universe, taking into account what constitutes 'positive' and 'negative' energy, is zero. Ingenious though these speculations are, they leave us with a nagging sense of dissatisfaction, a feeling that there is still something of fundamental importance to be discovered. If Wigner's view that consciousness is a phenomenon outside the scope of quantum mechanics is correct, we must at least look forward to its incorporation in a more comprehensive future physics. It is, however, possible that the Universe is of such a nature that it cannot contain a full description of itself, and even more likely that the human brain is not as yet sufficiently developed to be able to grasp ultimate cosmic truths.

Carbon-based life elsewhere?

It is certainly possible, even probable, that many other planets of stars in the observable Universe bear carbon-based organisms. The direction of evolution on different planets could well have been diverse, and the end-products dominating the biospheres might be disparate in form and capabilities. Even if, as Hoyle argues, life is common in the Universe in microbial form, and is seeding planets, the occurrence of critical situations and 'bifurcations', as envisaged by Prigogine, would lead to a multiplicity of products from sets of seemingly similar starting conditions. If Cairns-Smith is correct, many planets could have mineral organisms, but on only some of these planets would the conditions be right for the development of carbon-based organisms. On many of these, life might never pass beyond the microbial stage. The absence of direct evidence of the existence of extraterrestrial intelligent life, emphasized by many writers as an argument against its existence, is not really surprising. Even if it exists, the chances are that planets inhabited by intelligent beings able or wishing to communicate with each other are separated by vast distances that reduce the likelihood of interaction to a very low level. On the other hand, if highly superior beings able to survey our Earth were aware of our existence, it is quite likely that they would do so by means we might not detect. The remote possibility of detection of patterned radio messages, and possibly the detection of message-carrying laser beams, as recently suggested by Russian scientists, are for us the only currently available means of searching for evidence pointing to the existence of intelligent extraterrestrial beings. There is, however, always the possibility that we might be the first users of radio communication or inventors of lasers in the Universe, or that others might use means of communication unknown to us.

Functional arguments are sometimes used to support the idea that beings on other worlds might resemble ourselves. It is quite likely that in similar environments an essentially similar biochemistry would develop. The advantages of standing on two legs, of having eyes so placed that they can see afar, of having two arms, of having the brain in the head and so on are extolled, and there is some force in the argument that natural selection might tend to select similar successful systems. Eyes, for example, have apparently evolved by several different routes among Earth organisms. What is selected depends on the environment as well as on the constitution of organisms. A lobster or a centipede might have different views on what is good. However, lobsters and centipedes have not been known to express their views, but if they did we might fail to understand them through unfamiliarity with their communication systems. What humans consider to be the most highly intelligent animals – apes and humans – have the 'advantageous' features outlined. Recently, there has been much interest in the intelligence of dolphins and whales. It is apparent that in spite of their agility and precision of movement, and other admirable qualities, aquatic mammals devoid of limbs or other structures with which to manipulate materials would face difficult problems in initiation of technological developments. The structure and abilities of animals are of many kinds, and some planets might harbour organisms that could be described as intelligent but lacking technology. Failure to produce the equivalent of our mathematical geniuses would severely limit the ability of intelligent organisms to explore the Universe. The biosphere on the Earth has probably been greatly influenced by catastrophic events, unlikely to be duplicated in detail on other planets, and even approximate parallelism, except in basic chemistry, would appear to be unlikely.

Our 'creative', 'inventive' or 'intelligent' Universe may be more prone to produce large-scale biological – and psychological – diversity than uniformity, even if the basic ground-plan of organisms is essentially similar everywhere. The enormous diversity of life on the Earth shows us what is possible in one biosphere, but among the vast number of terrestrial species, humans alone have developed the necessary technology for space exploration. If the Earth should be the one seat of life, we can be fairly sure that, short of a catastrophic end, self-inflicted or as a result of a cosmic accident, life will in the coming ages spread from the Earth to other worlds. Even the possibility that we are the sole guardians of life should make us pause to consider the implied cosmic responsibility, for it would be we (or perhaps we should say Gaia, the whole Earth considered as a single, self-regulating organism) who had brought the idea of value into the Universe, or at least had made it explicit. A growing reverence for life is expressed in increasing concern for the quality of our terrestrial environment, and if we ever meaningfully reach for the stars we shall have to take care to prevent catastrophic disruption of other biospheres we might possibly encounter, or perturb with our instruments.

We must hope that in the context of the great adventure of space exploration, the petty jealousies and misunderstandings which have so far divided humans will, in a new and more enlightened perspective, be seen as intrinsically trivial, time-wasting, and potentially catastrophically destructive. The militarization of space, actively pursued by the so-called 'Great Powers', has nothing to commend it, and too many of the world's politicians seem to be mental and moral dwarfs. Religious fanaticism, an outmoded form of human activity, is another source of destructive ideas. In our Universe, we can be sure that there is so much to know and for us to discover, and tremendous excitement and pleasure to be derived from the process of discovery. A concerted international effort would lead to a rapid advance in our understanding of the Cosmos, and of the nature of living things and their significance in the Universe. It seems to us that this would be a much healthier mental exercise than the morbid preoccupation with destruction which afflicts so many of our contemporaries. We believe that there is cause for hope, and that advances in many fields of human intellectual and technological activity should in the coming century focus attention on what is truly progressive and constructive, so that the old order will pass away – always provided that the human species does not in the near future inflict on itself an orgy of destruction, something we can all work to prevent. This is not just a matter of opposing nuclear weapons, but of developing attitudes and patterns of interest, of discovery of the pleasures of understanding, or of trying to understand, the Universe of which we are organized, thinking components.

FURTHER INFORMATION
AND IDEAS

We hope that the books and articles listed below will provide the reader with more information on many of the points we have mentioned, and will draw attention to several avenues of approach to the problems of life's origin and nature that we have not been able to explore.

ASIMOV, I. (1980). *Extraterrestrial Civilizations*. Robson Books. (Pan Edition, 1981.)

BEATTY, J. K., O'LEARY, B. and CHALKIN, A. (1981). *The New Solar System*. Cambridge University Press.
A review of recent discoveries, with many good illustrations and an informative text.

BOHM, David. (1980). *Wholeness and the Implicate Order*. Routledge & Kegan Paul.
Fascinating ideas on matter, consciousness and quantum physics.

BRAND, Stewart. (1977). *Space Colonies*. A CoEvolution Book. Penguin Books.
Many experts write about and discuss ways in which humans might colonize space. Many illustrations.

BRIGGS, John P. and PEAT, F. David. (1984). *Looking Glass Universe*. Simon & Schuster.
A fascinating review of trends in thought leading to the establishment of new ideas in physics, chemistry, biology and neurobiology.

CALDER, Nigel. (1973). *The Life Game*. British Broadcasting Corporation.
Excellent summary of some of the new ideas in biology. Good illustrations.

CHURCHLAND, P. M. (1984). *Matter and Consciousness*. A Bradford Book. MIT Press.
Contains a thought-provoking discussion on the possible distribution of intelligence in the Universe.

CROWE, M. J. (1986). *The Extraterrestrial Life Debate 1750–1900. The Idea of a Plurality of Worlds from Kant to Lowell*. Cambridge University Press.
An extensively researched history, full of interesting information.

DAVIES, Paul. (1977). *Space and Time in the Modern Universe*. Cambridge University Press.

—— (1977). *The Physics of Time Asymmetry*. University of California Press.

—— (1979). *The Forces of Nature*. Cambridge University Press.

—— (1980). *Other Worlds*. A Touchstone book. Simon & Schuster.

—— (1983). *The Edge of Infinity*. Oxford University Press.

All of Paul Davies' books are excellent reading and valuable reviews of modern ideas in physics.

DAY, W. (1984). *Genesis on Planet Earth*. Yale University Press.

An excellent book on problems of the origin of life.

DOLE, S. H. (1964). *Habitable Planets for Man*. Blaisdell Publishing Co. New York.

DYSON, F. (1985). *Origins of Life*. Cambridge University Press.

Some type of cellular 'life' could have preceded genes.

FARLEY, J. (1977). *The Spontaneous Generation Controversy from Descartes to Oparin*. Johns Hopkins University Press.

A valuable history of ideas.

FERRIS, J. P. (1984). 'The Chemistry of Life's Origin'. *Chemistry & Engineering*. Aug. 27. pp. 22–35.

FOLSOME, C. E. (1979). *The Origin of Life: A Warm Little Pond*. Freeman.

Deals with many of the problems of life's origins. Suggests that protocells were formed from polymeric material in primæval ponds coincidently with small organic molecules. Discusses the evolution of the genetic apparatus.

GOLDSMITH, D. (ed.) (1980). *The Quest for Extraterrestrial Life. A Book of Readings*. University Science Books. Mill Valley, California.

A very useful collection of articles, extending from early times to the present.

GONICK, L. and WHEELIS, M. (1983). *The Cartoon Guide to Genetics*.

An entertaining and easy to follow guide to present-day ideas. Harper & Row.

HALDANE, J. B. S. (1965). 'Data Needed for a Blueprint of the First Organism'. In S. W. Fox (ed.), *The Origins of Prebiological Systems*. Academic Press.

HART, M. H. and ZUCKERMAN, B. (eds) (1982). *Extraterrestrials: Where Are They?* Pergamon Press.

HOROWITZ, N. H. (1986). *To Utopia and Back. The Search for Life in the Solar System*. Freeman.

This book was not available to us at the time of writing. From a review in *Nature*, we conclude that it will be obligatory reading for all interested in life in the universe.

JACOB, F. (1976). *The Logic of Life*. English translation of 1973 French Edition. Vintage Books. Random House.

A brilliant presentation of ideas in biology by a master scientist and Nobel laureate. He points out that an organization often has properties that do not exist at the level below, properties which can be *explained* by the properties of the components but not *deduced* from them. With living things, 'the deviations introduced by natural selection make any prediction impossible'. He concludes: 'Today the world is messages, codes and information. Tomorrow what analysis will break down our objects to reconstitute them in a new space? What new Russian Doll will emerge?'

MATSUNO, K., DOSE, K., HARADA, K. and ROHLFING, D. L. (eds) (1984). *Molecular Evolution and Protobiology*. Plenum Press.

MONOD, J. (1971). *Chance and Necessity*. Knopf. A Vintage Books edition appeared in 1972.

Original French Edition, Editions du Seuil, 1970. An authoritative review of the nature and ethical and philosophical implications of modern genetics, written by a leading scientist and Nobel laureate, who played a key role in fundamental biological research. The book concludes with the words, 'The ancient covenant is in pieces; man knows at last that he is alone in the universe's unfeeling immensity, out of which he emerged only by chance. His destiny is nowhere spelled out, nor is his duty. The kingdom above or the darkness below: it is for him to choose.'

NEWMAN, W. I. and SAGAN, CARL. (1981). 'Galactic Civilizations: Population Dynamics and Interstellar Diffusion'. *Icarus*. 46:293–327. For those who like to indulge in highly speculative thought. We have avoided any detailed discussion of galactic civilizations, as it seems to us that too many assumptions are often made. We need more hard information on such points as the frequency of planetary systems.

OBERG, J. E. (1981). *New Earths. Restructuring Earth and Other Planets*. New American Library. Meridian Book. New York.

ORO, J. (1980). 'Prebiological Synthesis of Organic Molecules and the Origin of Life'. In H. O. Halvorson and K. E. Van Holde (eds). *The Origins of Life and Evolution*. Alan Liss Inc. p. 47.

PAGELS, Heinz R. (1982). *The Cosmic Code*. Simon & Schuster. (Bantam Edition, 1983.) The author suggests that the Universe is a message written in a code, and that the scientist's job is to decipher the code. An excellent account of modern thinking in physics.

PONNAMPERUMA, C. (1972). *The Origins of Life*. Thames & Hudson.

—— (ed.) (1983). *Cosmochemistry and the Origin of Life*. D. Reidel, Dordrecht, Holland.

POWERS, Jonathan. (1982). *Philosophy and the New Physics*. Methuen. Looks at reasons for the different interpretations of modern physics, including those common in popular expositions. The author believes that profound conceptual misunderstandings can be 'masked by technical facility – important though that is'. A strongly recommended book.

PRICE, Charles C. (1974). *Synthesis of Life*. Dowden, Hutchinson and Ross Inc., Stroudsburg, Pennsylvania. A valuable collection of reprints of many of the basic papers in the field, with comments by the editor. A major reference source.

ROSNAY, Joel de. (1966). *Les Origines de la Vie, de l'Atome à la Cellule*. Editions du Seuil, Paris. Good general summary of the situation at that time.

SCHOPE, W. J. (1978). 'The Evolution of the Earliest Cells'. *Scientific American*. 239 No. 3: 110 ff. Sept. issue. Makes the point that for the first 3000 million years after the development of organisms on Earth, primitive organisms were the only living things. Has some excellent pictures of microfossils, and discusses the development of biochemical mechanisms.

SHKLOVSKII, I. S. and SAGAN, Carl. (1966). *Intelligent Life in the Universe*. A Delta Book. Dell Publishing Inc.

STROTHER, P. K. and BARGHOORN, E. S. (1980). 'Microspheres from the Swartkoppie Formation: A Review'. In H. O. Halvorson and K. E. Van Holde (eds). *The Origins of Life and Evolution*. Alan R. Liss Inc. p. 1. Looking for evidence of Earth's earliest inhabitants.

VIDAL, G. (1984). 'The Oldest Eukaryotic Cells'. *Scientific American*. 250 No. 2: 48 ff. Feb. issue.

Traces the origin of eukaryotic cells to unicellular plankton, about 1.4 thousand million years ago. Pictures of structures believed to be microfossils.

WALD, G. (1974). 'Fitness in the Universe: Choices and Necessities'. *Origins of Life*. 5: 7–27.

'We live in a Universe of chance, but not of accident.' Natural selection of a sort has extended even beyond the elements, to determine the properties of protons and electrons.

WEINBERG, S. (1977). *The First Three Minutes. A Modern View of the Origin of the Universe*. Bantam Books.

An excellent account of the 'standard' theory of the Big Bang and its consequences. It should be noted that new ideas developed since this book was written are leading to some modification of ideas on very early events: see for example an article on Alan Guth's 'Inflation' model, postulating an initial exceedingly rapid expansion lasting one million million million million millionth of a second, by which time the region that is now 10 billion light years across had reached a diameter of four inches. 'Standard cosmology' then began. See *Science 84*. Jan./Feb. issue. p. 44.

WEINTRAUB, P. (ed.). (1984). *The Omni Interviews*. Omni Press.

Contains interesting interviews with a number of scientists, including Cyril Ponnamperuma, Francis Crick and Ilya Progogine.

WOODRUFF, I., SULLIVAN III, T. and MIGHEL, K. J. (1985). 'A Milky Way Search Strategy for Extraterrestrial Intelligence'. *Icarus*. 60: 675–84.

YOKOO, H. and OSHIMA, T. (1979). 'Is Bacteriophage ΦX174 DNA a Message from an Extraterrestrial Intelligence?' *Icarus. 38*: 148–53.

Has someone, somewhere, used the structure of a nucleic acid to code a message that can replicate when it reaches a suitable world? A great idea but difficult to prove. Certainly, a message from another civilization might arrive in an unexpected form.

EVOLUTION OF EUKARYOTES. Recent work on molecular genetics and cell ultrastructure gives some support to the view that cells with membrane-bounded (eukaryotic) nuclei, but lacking mitochondria, engulfed bacteria to make the first mitochondria. Some present-day eukaryotic micro-organisms lack mitochondria, and may be living relics of early eukaryote evolution. See T. Cavalier-Smith, 'Eukaryotes with no Mitochondria', *Nature*, 26 March 1987, for an interesting discussion of this topic.

BIBLIOGRAPHY

ABELSON, P. H. (1956). 'Amino-Acids formed in "Primitive Atmospheres" '. *Science*. 124:935.

ADAMS, W. S. and DUNHAM, T. (1932). *Pub. Astron. Soc. Pacific.* 44:243.

ANDREW, S. P. S. (1979), 'Designing the First Organism'. *Chemistry in Britain*. 15:580–6. Nov.

ANTONIADI, E. M. (1916). 'Mem.B.A.A.'. 2:33–4. Also in *Mars*, 1975, p. 54. Translation from the original French edition by Patrick Moore. Keith Reid Ltd, Shaldon, Devon.

—— (1934). *La Planète Mercure et la Rotation des Satellites*, Paris. Translated by Patrick Moore as *The Planet Mercury*. Keith Reid Ltd, Shaldon, Devon, 1974.

ARRHENIUS, Svante. (1909). *The Life of the Universe*. Harper & Brothers.

—— (1918). *The Destinies of the Stars*. Putnam. London.

BAKER, B. L. (1970–1). 'Review of Organic Matter in the Orgueil Meteorite'. In *Space Life Science*. 2:472–97.

BALDWIN, R. B. (1963). *The Measure of the Moon*. University of Chicago Press.

BARGHOORN, E. S. (1971). 'The Oldest Fossils'. *Scientific American*. May. Reprinted in *Life: Origin and Evolution*. Freeman. 1979.

BASTIAN, H. C. (1905). *The Nature and Origin of Living Matter*. Fisher Unwin.

BAUGHER, Joseph F. (1985). *On Civilized Stars*. Prentice-Hall.

BERNAL, J. D. (1951). *The Physical Basis of Life*. Routledge & Kegan Paul.

—— (1967). *The Origin of Life*. Weidenfeld & Nicolson.

BRASIER, M. D. (1979). 'The Early Fossil Record'. *Chemistry in Britain*. 15:588–92. Nov.

BRIGGS, M. H. (1960). 'The Colouring Matter and Radio Emissions of Jupiter'. *The Observatory*. 80:159.

BUHL, D. (1974). 'Galactic Clouds of Organic Molecules'. *Origins of Life*. 5:29–40.

BUNGENBERG DE JONG, H. C. (1931–2). *Protoplasma* 15: 110. And in Oparin (1957).

CAIRNS-SMITH, A. G. (1971). *The Life Puzzle*. University of Toronto Press.

—— (1979). 'Organisms of the First Kind'. *Chemistry in Britain*. 15:576–9. Nov.

—— (1985). *Seven Clues to the Origin of Life*. CUP.

CALVIN, M. (1959). 'Evolution of Enzymes and the Photosynthetic Apparatus'. *Science*. 130: 1170.

—— (1969). *Chemical Evolution. Molecular Evolution Towards the Origin of Living Systems on the Earth and Elsewhere*. Oxford, Clarendon Press.

CARR, M. H. (1981). *The Surface of Mars*. Yale University Press.

CLAUS, G. AND NAGY, B. (1961). 'A Microbiological Examination of some Carbonaceous Chondrites'. *Nature*. 192:594–6.

COLE, K. C. (1985). 'Much Ado about Nothing'. (Article on the vacuum.) *Discover*. June. 76–80.

CRICK, Francis. (1982). *Life Itself, its Origin and Nature*. Touchstone book. Simon & Schuster.

CROWTHER, J. G. (1955). In *Science Unfolds the Future*. ch. vi. Fredrick Muller.

CRUTCHFIELD, J. P., FARMER, J. D., PACKARD, N. H. and SHAW, R. S. (1986). 'Chaos. There is Order in Chaos: Randomness has an Underlying Geometric Form'. *Scientific American*. 255:6 pp. 38–49.

DAVIES, Paul. (1982). *The Accidental Universe*. Cambridge University Press.

—— (1983). *God and the New Physics*. Dent.

DEAMER, D. W. and BURCHFIELD, G. L. (1982). 'Encapsulation of Macromolecules by Lipid Vesicles under Simulated Prebiotic Conditions'. *J. Mol. Evol.* 18:203.

DENBIGH, K. G. (1975). *An Inventive Universe*. George Braziller.

De VAUCOULEURS, G. (1954). *Physics of the Planet Mars*. London. p. 126.

DILLON, L. S. (1978). *The Genetic Mechanism and the Origin of Life*. Plenum Press.

DOBELL, C. (1932). *Anthony van Leeuwenhoek and his Little Animals*. Staples Press, London.

DOLLFUS, A. (1951). *Comptes Rendus*. 1066, p. 232.

DOSE, K., FOX, G. A., DEBORIN, G. A. and PAVLOVSKAYA, T. E. (1974). *The Origin of Life and Evolutionary Biochemistry*. Plenum Press.

EDDINGTON, A. S. (1935). *New Pathways in Science*. Cambridge University Press.

EIGEN, Manfred. (1971). 'Self-organization of Matter and the Evolution of Biological Molecules'. *Naturwissenschaften*. 50(10):465–523. Reprinted in Charles C. Price (ed.), *Synthesis of Life*. Hutchinson & Ross Inc. Dowden.

—— and WINKLER, R. (1982). *Laws of the Game*. Allen Lane. (Translation of the 1975 German edition.)

EYRING, H. and JOHNSON, F. H. (1957). 'The Critical Complex Theory of Biogenesis'. In *Influence of Temperature on Biological Systems*. Am. Physiol. Soc., Washington. p. 1.

FEINBERG, G. and SHAPIRO, Robert. (1980). *Life Beyond Earth*. William Morrow.

FITCH, F. W. and ANDERS, E. (1963). 'Observations on the Nature of the "Organized Elements" in Carbonaceous Chondrites'. *Ann. N.Y. Acad. Sci.* 108:495–513.

FOLSOME, C. E. (1979). *Life: Origin & Evolution*. Scientific American articles by various authors.

—— and MOROWITZ, H. J. (1968–9). 'Prebiological Membranes: Synthesis and Properties'. *Space Life Sciences*. 1:538–544.

FOX, S. W. (1956). 'Evolution of Protein Molecules and Thermal Synthesis of Biochemical Substances'. *American Scientist*. 44:347.

—— and DOSE, K. (1977). *Molecular Evolution and the Origin of Life*. Marcel Dekker.

——, HARADA, K. and KENDRICK, J. (1959). 'Production of Spherules from Synthetic Proteinoid and Hot Water'. *Science*. 129:1221.

FRIEDMANN, E. I. (1980). 'Endolithic Microbial Life in Hot and Cold Deserts'. in *Origins of Life*. 10:223–235. And see: 'Living in Rock and Lichen it', by Marcia BARTUSIAK. *Science. 83*, April issue, p. 74.

GARRISON, W. M. and MORRISON, D. C., HAMILTON, J. G., BENSON, A. A. and CALVIN, M. (1951). 'Reduction of Carbon Dioxide in Aqueous Solutions by Ionizing Radiations'. *Science. 114*:416.

GILLETT, S. L. (1985). 'The Rise and Fall of the Early Reducing Atmosphere'. *Astronomy. 13*:7. July, p. 66.

GOLDACRE, R. J. (1958). 'Surface Films, their Collapse on Compression, the Shapes and Sizes of Cells and the Origin of Life'. In J. F. Danielli, K. G. A. Pankhurst and A. C. Reddiford (eds). *Surface Phenomena in Chemistry and Biology*. Pergamon Press.

GOLDANSKII, V. I. (1986). 'Quantum Chemical Reactions in the Deep Cold'. *Scientific American*. Feb. p. 46.

GOLDSCHMIDT, V. M. (1952). 'Geochemical Aspects of the Origin of Complex Organic Molecules on the Earth, as Precursors to Organic Life'. *New Biology. 12*:97. Penguin.

HALDANE, J. B. S. (1929). 'The Origin of Life'. *Rationalist Annual*.

—— (1954). 'The Origins of Life'. *New Biology. 16*:12–27. Penguin.

HASSELSTROM, T., HENRY, M. C. and MURR, B. (1957). 'Synthesis of Amino-acids by Beta Radiation'. *Science. 125*:350.

HORNE, R. A. (1971). 'On the Unlikelihood of Non-aqueous Biosystems'. *Space Life Sciences. 3*:34.

HOROWITZ, N. H. (1957). 'On Defining "Life" '. In *The Origin of Life on the Earth*, p. 106. Pergamon, London.

—— (1977). 'The Search for Life on Mars'. *Scientific American*, Nov. issue. Reprinted in *Life: Origin and Evolution*. Freeman. 1979.

HOYLE, Fred. (1955). *Frontiers of Astronomy*. Heinemann, London.

—— (1983). *The Intelligent Universe*. Michael Joseph.

—— (1986). 'Halley's Comet and Others; the Bacterial Star Shells'. *Proc. Royal Soc. Med. 79*:12, pp. 691–3.

—— and WICKRAMASINGHE, N. C. (1978). *Lifecloud*. Sphere.

—— (1979). *Diseases from Space*. Dent.

—— (1981). *Evolution from Space*. Dent.

—— (1985). *Living Comets*. University of Cardiff Press.

HUNT, G. E. and MOORE, Patrick. (1982). *The Planet Venus*. Faber & Faber. London.

—— (1987). *Uranus*. Cambridge University Press.

HUNTEN, D. M. (ed.) (1983). *Venus*. University of Arizona Press. Tucson.

JEANS, J. H. (1930). *The Universe Around Us*. Cambridge University Press.

KHARE, B. H. and SAGAN, Carl. (1981). 'Organic Solids Produced by Electrical Discharge in Reducing Atmospheres: Tholin Molecular Analysis'. *Icarus. 48*:290–297.

KIEFFER, H. H. and PALLUCONI, F. D. (1979). 'The Climate of the Northern Polar Cap. [Mars]'. *NASA CP-2072*, pp. 45–6.

KLEIN, H. P. (1978–9). 'The Viking Biological Investigations. Review and Status'. *Origins of Life. 9*:157–60.

KRALLENBERG, R. *et al.* (1980). See *J. Geophys. Res., 85*:8059 ff. And see KRALLENBERG, R. and HUNTEN, D. M. (1980). *J. Geophys. Res., 85*:8039 ff.

HUXLEY, T. H. (1868). 'On the Physical Basis of Life'. Reprinted in *Lectures and Essays*, T. H. Huxley. Thinker's Library. Watts & Co., London, 1931.

KONDRATYEV, K. Y. and HUNT, G. E. (1982). *Weather and Climate on the Planets*. Pergamon. Ch. 4.

KOZYREV, N. A. (1954). See *Publ. Crimean Astrophys. Obs.*, 12:177.

LANGONE, J. (1983). 'Cyril Ponnamperuma: Meteorites and the Stuff of Life'. *Discover*. Nov. pp. 51–60.

LEDERBERG, J. (1960). 'Exobiology: Approaches to Life Beyond the Earth'. *Science*. 132:393.

LIND, S. C. and BARDWELL, D. C. (1926). *J. Am. Chem. Soc.* 48:2335.

LOVELOCK, J. E. (1979). *Gaia. A New Look at Life on Earth*. Oxford University Press.

—— (1986). 'Gaia: The World as Living Organism'. *New Scientist*. 18 Dec. 1986, pp. 25–8.

LOWELL, P. (1906). *Mars and its Canals*. Macmillan, New York. p. 376.

LUNAN, D. (1979). *New Worlds for Old*. William Morrow.

LURIA, S. E. (1973). *Life the Unfinished Experiment*. Scribner's.

MARGULIS, Lynn. (1971). 'Symbiosis and Evolution'. *Scientific American*. Aug.

—— (1981). *Symbiosis in Cell Evolution, Life and its Environment on the Early Earth*. Freeman.

—— and SAGAN, Dorion. (1985). 'L'Origine des Cellules Eukaryotes'. *La Recherche*. Paris. Feb. 1985.

MAYNARD-SMITH, J. (1986). *Problems of Biology*. Oxford University Press.

MENZEL, D. H. and WHIPPLE, F. L. (1955). See *Pub. Astron. Soc. Pacific*. 67:161 ff.

MILLER, S. L. (1953). 'A Production of Amino-Acids under Possible Primitive Earth Conditions'. *Science*. 117:528.

—— and UREY, H. C. (1955). 'Organic Compound Synthesis on the Primitive Earth'. *Science*. 130:245.

—— (1959). 'Formation of Organic Compounds on the Primitive Earth'. In *The Origin of Life on the Earth*. Pergamon Press.

MOORE, B. (1913). *Origin and Nature of Life*. Home University Library of Modern Knowledge, London, Williams & Norgate.

MOORE, Patrick. (1976). *Guide to the Moon*. Lutterworth, London. Norton.

—— (1977a). *Guide to Mars*. Lutterworth Press, Guildford.

—— (1977b). 'Requiem for the Canals'. *J. Br. Astron. Assoc.*, 87:6. pp. 589–93.

—— (1984). 'The Mapping of Mars'. Presidential address to the British Astronomical Association, 1983. *J.B.A.A. Vol. 94*. pp. 45–54.

—— (1985). 'The Mapping of Venus'. Presidential Address to the British Astronomical Association, 1984. *J.B.A.A. 95*: pp. 50–61.

MORRISON, D. (1982). *Voyages to Saturn*. NASA. Washington, D.C.

NAGY, B. and CLAUS, G. (1963). 'Mineralized Microstructures in Carbonaceous Meteorites'. In *Organic Geochemistry*, U. Colombo and G. D. Hobson (eds). Pergamon Press. pp. 109–14.

NICOLSON, I. K. (1978). *The Road to the Stars*. Westbridge Books, London.

NODA, H. (1978). *Origin of Life*. Proceedings of the 5th. ICOL Meeting, Kyoto, Japan. Centre for Academic Publications, Japan. Japan Scientific Societies Press.

NURSALL, J. R. (1959a). 'Oxygen as a Prerequisite to the Origin of the Metazoa'. *Nature*. 183:1170.

—— (1959b). 'The Origin of the Metazoa'. *Transactions of the Royal Society of Canada. LIII*. Series III. June. Section 5. p. 1.

OLSON, E. C. (1985). 'Intelligent Life in Space'. *Astronomy. 13*:7. July. p. 7.
OPARIN, A. I. (1957). *The Origin of Life on the Earth*. Oliver & Boyd.
—— (1961). *Life: Its Nature and Origin*. Oliver & Boyd.
—— (1964). *The Chemical Origin of Life*. Charles C. Thomas.
—— (1968). *Genesis and Evolutionary Development of Life*. Academic Press.
ÖPIK, E. J. (1950). *Irish Astronomical Journal. 1*. p. 22.
ORGEL, E. L. (1973). *The Origins of Life*. Wiley.
OWEN, T. (*et al.*) (1977). *J. Geophys. Res., 82*, pp. 635–9.
PASCHKE, R. and CHANG, R. W. H. and YOUNG, D. (1957). 'Probable Role of Gamma-Irradiation in the Origin of Life'. *Science. 125*:881.
PAULING, L. (1957). See *The Origin of Life on the Earth*. I.U.B. Symposium Series. Pergamon. p. 119.
PAVLOVSKAYA, T. E. and PASYNSKII, A. G. (1959). 'The Original Formation of Amino-Acids under the Action of Ultra-Violet Rays and Electrical Discharges'. In *The Origin of Life on the Earth*. Pergamon. p. 151.
PERRET, J. (1952). 'Biochemistry and Bacteria', *New Biology. 12*:69. Penguin.
PIRIE, N. W. (1937). 'The Meaninglessness of the Terms "Life" and "Living" '. In *Perspectives in Biochemistry*. Cambridge University Press.
—— (1954). 'On Making and Recognizing Life'. *New Biology. 16*:41. Penguin.
PONNAMPERUMA, C. (ed.) (1981). *Comets and the Origin of Life*. D. Reidel.
POPPER, Karl R. (1972). In *Objective Knowledge*. Clarendon Press. Oxford. See 'On the Theory of the Objective Mind'. pp. 153–90.
—— (1982). In *Quantum Theory and the Schism in Physics*. Hutchinson. See 'Quantum Mechanics Without "the Observer" ', pp. 35–95.
POUNDSTONE, W. (1985). *The Recursive Universe. Cosmic Complexity and the Limits of Scientific Knowledge*. Contemporary Books Inc. Chicago.
PRIGOGINE, Ilya and STENGERS, Isabelle. (1984). *Order out of Chaos*. Bantam Books.
PRINGLE, J. W. S. (1953). 'The Origin of Life'. *Symp. Soc. Exptl. Biol., 7*:1.
—— (1954). 'The Evolution of Living Matter'. *New Biology. 16*:54. Penguin.
REEVES, Hubert. (1985). *Atoms of Silence*. MIT Press.
SCHÄFER, E. A. (1912). *Rep. Brit. Ass. p.* p. 3.
SCHRÖDINGER, E. (1944). *What is Life?* Cambridge University Press.
SCHULTZ, O. H. (1977). *Moon Morphology*. University of Texas Press.
SOKOLOV, V. (1937). English Abstract, Abstracts of Papers. *17th. International Geological Congress*, Moscow.
SPIEGELMAN, S. (1970). 'Extracellular Evolution of Replicating Molecules'. *The Neurosciences: Second Study Program*. F. O. Schmidt (ed.) pp. 927–45. Rockefeller University Press. Reprinted in *Synthesis of Life*. Charles C. Price (ed.). Dowden, Hutchinson Ross Inc. 1984.
TOMBAUGH, C. and MOORE, Patrick. (1980). *Out of the Darkness: the Planet Pluto*. Stackpole Books. Harrisburg. Pennsylvania and Lutterworth Press. London.
TYNDALL, J. (1876). 'Vitality'. In *Fragments of Science*. Longmans, Green & Co. London. 2nd edn. p. 459. First written in 1865.
VAN DE KAMP, P. (1987). In *Encyclopaedia of Astronomy*. Mitchell Beazley. London.
VINOGRADOV, A. P. (1959). 'The Origin of the Biosphere'. In *The Origin of Life on the Earth*. Pergamon. p. 23.
WATSON, J. D. (1968). *The Double Helix*. Atheneum Publishers.
WHEELER, J. A. (1982). 'Bohr, Einstein and the Strange Lesson of the Quantum'. In *Mind in Nature*. Nobel Conference XVII. Harper & Row. p. 1.

WIGNER, E. (1982). 'The Limitations and Validity of Present Day Physics'. In *Mind in Nature*. Nobel Conference XVII. Harper & Row. p. 118.

WILDT, R. (1937). See *Astrophysical Jnl.*, *86*:321.

WOESE, C. R. (1980). 'An Alternative to the Oparin View of the Primeval Sequence'. In *The Origins of Life and Evolution*. H. O. Halvorson and K. E. Van Holde (eds). Alan R. Liss Inc. p. 65–76.

INDEX